Mendeleyev's Dream

By the Same Author

Science

Pythagoras and His Theorem
Archimedes and the Fulcrum
Galileo and the Solar System
Newton and Gravity
Darwin and Evolution
Curie and Radioactivity
Einstein and Relativity
Bohr and Quantum Theory
Turing and the Computer
Oppenheimer and the Bomb
Crick, Watson and DNA
Hawking and Black Holes

Novels

Pass by the Sea
A Season in Abyssinia
One Man's War
Vaslav
The Adventures of Spiro

Mendeleyev's Dream

The Quest for the Elements

Paul Strathern

Thomas Dunne Books
St. Martin's Press New York

THOMAS DUNNE BOOKS.
An imprint of St. Martin's Press.

www.stmartins.com

ISBN 0-312-26204-3

First published in Great Britain by Hamish Hamilton Ltd, The Penguin Group

First U.S. Edition: January 2001

10 9 8 7 6 5 4 3 2 1

The chemists are a strange class of mortals, impelled by an almost maniacal impulse to seek their pleasures amongst smoke and vapour, soot and flames, poisons and poverty, yet amongst all these evils I seem to live so sweetly that I would rather die than change places with the King of Persia.

Johann Joachim Becher, *Physica Subterranea* (1667)

The elementes turned in to themselues, like as whan one tune is chaunged vpon an instrument of musick.

Coverdale Bible, Wisdom, xix, 18

Contents

List of Illustrations

Credits: AKG London, 2, 3, 4, 6, 9; AKG London/Erich Lessing, 7; Editions Seghers, Paris, 1; Metropolitan Museum of Art, New York (Purchase, Mr. and Mrs. Charles Wrightsman Gift, in honor of Everett Fahy, 1977), 8

Prologue

There is a faded photograph of the Russian chemist Dmitri Mendeleyev at work in St Petersburg, taken some time in the late nineteenth century. It shows a gnomic figure seated at a vast littered desk. Mendeleyev looks not unlike some Siberian shaman transposed to surroundings recognizable as the habitat of the shaman's modern successor: the genius-professor's study. He has a long white unkempt beard, which ends in three distinct points – indicative of obsessively combing fingers during periods of absent-minded pondering. His unkempt white hair trails to his shoulders. Mendeleyev was in the habit of having his hair cut annually. At the onset of the warmer spring weather he would summon a local shepherd, who would attend to this matter with sheep shears. This is the famous shock of hair which once caused Mendeleyev to be described by the Scottish chemist Sir William Ramsay as 'a peculiar foreigner, every hair of whose head acted in independence of every other'. Ramsay supposed Mendeleyev was from Siberia, judging him to be 'a Kalmuch, or one of those outlandish creatures'.

In the photograph Mendeleyev is concentrating over a piece of paper, writing with dip pen held between the very ends of his long fingers. The rest of the wide and irregular desktop is confusion. Pieces of paper upon pieces of paper, a mug on a saucer, a number of instruments of indeterminable purpose and, on shelves beneath the desk, haphazardly stacked files of scientific papers.

Mendeleyev at his desk

At Mendeleyev's back there is a bookcase containing three surprisingly ordered rows of bound volumes. In the midst of these a large labelled key hangs directly above his head – like a kind of scientific halo, or exclamation mark. (Eureka!) And above the bookcase the patterned wallpaper of the period is covered with uneven rows of shadowy framed pictures, portraits of the great scientists of the past. From the upper dimness Galileo, Descartes, Newton and Faraday gaze down at the shock-haired figure scribbling away amidst the enveloping disorder.

In 1869 Mendeleyev was puzzling over the problem of the chemical elements. They were the alphabet out of which the language of the universe was composed. By this time sixty-three different chemical elements had been discovered. These ranged from copper and gold, which had been known since prehistoric times, to rubidium, which had recently been detected in the atmosphere of the sun. It was known that every one of these elements consisted of different atoms, and that the atoms of each element had their own unique properties. However, some elements had been found to possess vaguely similar properties, enabling them to be classified together in groups.

The atoms which made up the different elements were also known to have different atomic weights. The lightest element was hydrogen, with an atomic weight of 1. The heaviest known element, lead, was thought to have an atomic weight of 207. This meant that the elements could be listed in linear form, according to their ascending atomic weights. Or they could be assembled in groups with similar properties. Several scientists had begun to suspect that there was a link between these two methods of classification – some hidden structure upon which all the elements were based.

In the previous decade Darwin had discovered that all life forms progressed by evolution. And two centuries earlier Newton had discovered that the universe worked according to gravity. The chemical elements were the linchpin between the two. The discovery of a structure here would do for chemistry what Newton had done for physics and Darwin for biology. It would reveal the blueprint of the universe.

Mendeleyev was aware of the significance of his search.

This could be the first step towards uncovering, in future centuries, the ultimate secret of matter, the pattern upon which life itself was based, and perhaps even the origins of the universe.

Seated at his desk beneath the portraits of the philosophers and the physicists, Mendeleyev continued to ponder this seemingly insoluble problem. The elements had different weights. And they had different properties. You could number them and you could group them. Somehow there just had to be a link between these two patterns.

Mendeleyev was professor of chemistry at St Petersburg University, and was renowned for his encyclopedic knowledge of the elements. He knew them as a headmaster knows his pupils – the volatile bad mixers, the bullies, those easily led, the mysterious underachievers and the dangerous ones you have to watch. Yet try though he might, he could still discern no overall guiding principle amidst this whirligig of characteristics. There had to be one somewhere. The scientific universe couldn't just be based on some random collection of unique particles. That would be unscientific.

Mendeleyev's colleague A. A. Inostrantzev, who called in to see him on 17 February 1869, has left an account of this meeting. The account is somewhat imaginative and coloured by hindsight, but supplies some background details. Mendeleyev himself has also contributed various (somewhat differing) accounts of what he was doing and thinking at that time.

Apparently, for three days, almost without cease, Mendeleyev had been racking his brains over the problem of the elements. But he was aware that time was running out. That very day he was scheduled to catch the morning train from Moscow Station to his small country estate in the province of Tver. The professor of chemistry had a meeting with the Voluntary Economic Cooperative of Tver. He was due to address a delegation of local cheese-makers, to advise them on production methods, followed by a three-day tour of inspection of local farms. His wooden travelling trunk had already been packed and was standing in the hall. Through the window of his study the horse-drawn sleigh could be seen waiting in the street, the swaddled driver stamping in the snow, the forlorn nag wheezing white plumes into the frigid air.

But the servants wouldn't have dared to disturb Mendeleyev. His temper was notorious. On occasions he had been known to fly into such a rage that he would literally dance with fury. But what would happen if he missed the train?

The Russian psychologist of scientific discovery B. M. Kedrov, and other commentators on Mendeleyev, have speculated that the pressing thought of the train he was due to catch had an effect on Mendeleyev's mind. It may have been this which induced in him a glide of daydreaming inspiration . . . On the long journeys from St Petersburg to Tver, Mendeleyev would frequently while away the time playing patience. After setting up his wooden trunk between his knees, he would lay the pack face downwards. As the silver birches, the lakes and wooded hills slid past beyond the window, he would begin turning the cards three by three. When he came to the aces he would remove them one after the other, placing each suit in line at the top of the trunk: hearts, spades, diamonds, clubs. Then he would continue turning the cards – and one by one they would appear. King on the hearts, Queen on the hearts, King on the diamonds, Jack on the hearts . . . Slowly the suits would begin descending down the trunk. Ten, nine, eight . . . Suits, descending numbers. It was exactly the same as the elements with their groups and ordered atomic numbers!

At some point during the morning Mendeleyev must have called from his study door, and ordered one of the servants to dismiss the waiting horse-drawn sleigh. He was to inform the driver that the professor would now be catching the afternoon train.

Mendeleyev returned to his desk and began searching amongst the drawers. Eventually he pulled out a pile of white cards. (As he did so, he might have been aware of the jingling bells of the horse-drawn sleigh receding into the misty snow-padded distance.) One by one Mendeleyev began writing on the blank white surfaces of the cards. First he printed the chemical symbol of an element, then its atomic weight and finally a short list of its characteristic properties. When he had filled sixty-three cards he spread them out face upwards over the desk.

He began staring at the cards, ruminatively combing his fingers through the ends of his beard. One long moment stretched into minutes on end as he concentrated on the sea of cards before him, his mind following trails of thought and half-thought, lost to the world. But still he could discern no overall pattern.

An hour or so later he decided to try a different tack. He gathered up the cards and began laying them out in groups. Another timeless hour; his eyelids were beginning to flutter now with exhaustion. Finally, in despair, he decided to try the obvious course, laying out the cards in ascending order of their atomic weights. But this couldn't possibly lead to anything. Everyone else had tried it. And besides, weight was just a physical property. What he was looking for was a pattern which linked the chemical properties. By now his head was beginning to nod, falling forward over the cards as he checked himself, on the brink of sleep. It seems he became aware of the horse-drawn sleigh waiting outside the window. Was it still there? Or had it come back? Already? Was it time to catch the afternoon train? That was the last one, he couldn't possibly miss it.

As Mendeleyev's eyes ran once more along the line of ascending atomic weights, he suddenly noticed something that quickened his pulse. Certain similar properties seemed to repeat in the elements, at what appeared to be regular numerical intervals. Here was something! But what? A few of the intervals began with a certain regularity, but then the pattern just seemed to peter out. Despite this, Mendeleyev soon became convinced that he was on the brink of a major breakthrough. There was a definite pattern there somewhere, but he just couldn't quite grasp it . . . Momentarily overcome by exhaustion, Mendeleyev leaned forward, resting his shaggy head on his arms. Almost immediately he fell asleep, and had a dream.

I

In the Beginning

In order to understand the problem which Mendeleyev was attempting to solve, it is necessary to go back to the very origin of scientific thought. This seminal event in human evolution can be pinpointed to an exact time and an exact place – 2,500 years earlier in ancient Greece. The first appearance of genuine scientific thinking is traditionally credited to Thales, who lived in the sixth century BC in the Greek city of Miletus on the shore of Ionia (now south-western Turkey).

At this period Greek-speaking people occupied the scattered islands and isolated plains along the rocky shoreline of the entire Aegean Sea. It was this fragmented geography which partly accounted for why the Greek world had remained a collection of often quarrelsome city and island states, with a network of trading colonies all over the Mediterranean region. During this period, as throughout the time of their greatness, the ancient Greeks were united only by their language. Indicatively, the territory they occupied had no overall name. It was the Romans who first applied the word 'Graecia' to the land inhabited by the Greeks, who referred themselves as Hellenes (after Hellen, the chieftain of an obscure ancient tribe from Thessaly). It has been argued that the ancient Greeks weren't even a racial entity, but consisted of several different types sharing a common language, with a common culture and religion.

Why humanity should first have started to think in an essentially rational scientific manner in this particular region remains something of a mystery. Well over a millennium beforehand both the Babylonians and the ancient Egyptians had developed superior civilizations capable of highly advanced thought. Standing on the high terraces of their ziggurats the Babylonian astrologers had observed the night sky, discerning patterns amongst the stars, plotting their movements across the heavens. After the annual floods receded from the Nile valley leaving a sea of mud, the scribes of Egypt would recalculate the exact area of each individual plot of land, juggling mathematical units as small as 1/300. Such advanced observation and measurement had been the domain of the priestly castes in both Babylon and Egypt. These technical abilities, and any theoretial speculation they provoked, were a part of religious practice. Their subtlety had blended into the subtlety of the accompanying theology – resulting in such things as number magic, and the idea that the movements of the stars mirrored earthly fates. Such superstitions persist to this day in the form of 'lucky' numbers and 'personal star signs'.

By comparison, the ancient Greek religion, of ill-behaved gods roistering and philandering on Mount Olympus, was a joke. During the dark age which followed the collapse of Mycenaean civilization, the Greek religion had remained in its infantile stage of development; no serious quasi-scientific thinking could possibly be attached to such comic-book antics.

But this was the crucial point. When the first stirring of scientific curiosity took place in ancient Greece, it was not linked to any religion. It did not have to conform to the dictates of some well-entrenched theology, nor was it prompted to take off into some metaphysical cloud-cuckoo-land. It was entirely free – restrained by nothing but reason and the actuality of the world it confronted.

According to legend, Thales is said to have enjoyed walking in the hills above Miletus. We can picture the scene. Laid out below him in the glaring sunlight, the harbour city with its pristine marble columns and grid pattern of streets, exact and fragile as a miniature on an eggshell. The bays and capes of the Asian mainland stretching away to the north, the wide expanse of the still blue Aegean, distant islands shadowy in the heat haze. Perhaps a slack-sailed trading ship, becalmed in the gulf on its way back from one of the city's colonies on the Nile delta, the Black Sea, or Sicily – or even a voyage as far afield as the Spanish silver mines beyond the Pillars of Hercules (Gibraltar).

As Thales was walking along the hillside path he noticed some rocks which contained fossils of what were unmistakably seashells. He realized that these hills had once been part of the sea. This led him to surmise that originally the world must have consisted entirely of water. He concluded that water was the fundamental element from which all things derived. As a result, Thales of Miletus is generally regarded as the first philosopher. His was the first known example of truly scientific thinking.

In its early days philosophy included science – which became known as 'natural philosophy'. Thales' thinking was scientific because it could provide evidence for its conclusions. And it was philosophy because it used reason to reach these conclusions: there was no appeal to the gods or mysterious metaphysical forces. The argument was conducted entirely within the realms of this world, from which evidence could be gathered to prove or disprove its conclusions.

We know little for certain about Thales of Miletus. He is said to have predicted an eclipse of the sun which is now known to have taken place in 585 BC. This is the only real evidence we have for when he lived. According to one story he fell off a hill while studying the sky: an emblematic image of the philosopher to this day. But Thales was no fool. When asked why, if he was so clever, he remained poor, he replied that getting rich was easy. And proved it. Realizing that the coming olive crop would be good, he hired all the local olive presses. During the bumper olive harvest his monopoly enabled him to charge whatever he wished.

It is impossible to exaggerate the effect of the new philosophical way of thinking which is accredited to Thales. The knowledge of Western civilization is based upon it. Looking back, we can see that right from the start this new way of thinking contained certain basic assumptions. These were to determine (and over two and a half millennia later continue to determine) both the form and the content of our knowledge. These were the assumptions which were to underpin all subsequent scientific thinking. Thales asked the question, 'Why do things happen the way they do?' In answering this question he assumed that the answer must be in terms of the basic matter of which the world is made. He also assumed that there is an underlying unity beneath the diversity of the world. But perhaps most significantly of all, he assumed that there is an answer to this question. And that this answer can be given in the form of a testable theory – a word which derives from the Greek for 'to look at, contemplate, or speculate'.

We know from anecdotal evidence that Thales arrived at his theory after seeing some seashell fossils high above the contemporary sea level. But his speculations probably went deeper than this. He must have seen the mist rising from the Anatolian hills to become clouds, and have observed the rain falling from clouds in storms out over the Aegean. Land becoming damp air, which in turn became water. Just a couple of miles north of Miletus, a large river meanders over the wide plain to the sea. (This is in fact the ancient River Meander, from which our word derives.) Thales would have observed the river slowly silting up: the water becoming muddy earth. He would have visited the springs on the nearby hillside: the earth becoming water again. It takes little imagination now to see how Thales conceived of the idea that all is water. Yet the first person to step into the unknown with this idea must have been a giant of the imagination.

Curiously, this was not the only major step in the evolution of human thought which took place during the sixth century BC. Quite independently, in other parts of the globe, humanity was taking several major steps which would affect the entire course of its development. China witnessed the arrival of Confucius and Lao-Tzu (the founder of Taoism, the rival of Confucianism), Buddha began preaching in India, and in Persia the fire-worshipper Zarathustra founded Zoroastrianism (which was to have a major influence on both Judaism and Islam). Meanwhile the Mediterranean region was witnessing more than just the advent of the first philosophers in Ionia. By the end of the century Pythagoras was living at the other end of the Greek world in southern Italy. Pythagoras' religious teachings were to contribute significantly to the non-Judaic element in Christianity (several New Testament parables originate from Pythagorean sources). Likewise, the influence of Pythagoras' number religion is recognizable in musical theory and in modern science's belief that the ultimate workings of the universe can be described in terms of number. The direction of both Eastern and Western civilizations was set by events which took place during the sixth century BC.

The most important, for the Western world, was certainly the development attributed to Thales of Miletus, the first philosopher-scientist. His theory that the world had developed from one element (water) was just the beginning. This idea, once conceived, was quickly developed by Thales' pupils in Miletus – the philosophers known as the Milesian school. One of these was Anaximenes, who spotted the weak link in Thales' argument. If everything had originally been water, what accounted for the present diversity of the world? How had water become everything? Anaximenes argued that the fundamental element wasn't water, it was air. The world was surrounded by air, which became compressed the closer it came to the centre. As it became compressed, it turned into water; when water was further compressed it became earth; more compressed still, it became stone. Everything was air, in a more or less condensed state.

This was a significant development of Thales' idea. Here was the first attempt to explain the diversity of the world – the first attempt to explain qualitative difference in terms of quantitative difference.

A new way of thinking had been discovered. The great philosophical debate (which remains unresolved to this day) had started. Now everyone could join in. How Thales reacted to all this remains unrecorded. He may have thought scientifically; whether he expected to be contradicted scientifically is another matter, but he was. Anaximenes appears to have been a young man around the time that Thales died, and he passed on the story of Thales' death. In a letter to Pythagoras he wrote: 'Thales has met an unkind fate in his old age. As was his custom, he left his house at night with his maidservant to view the stars. But while he was gazing at the heavens he forgot where he was and stepped off a steep slope.' Anaximenes foresaw an even worse fate for himself. In a later letter to Pythagoras, he bemoans: 'How can I think of studying the stars when I am faced with the prospect of slaughter or slavery?' By this time the scattered city states of the Greek world were under threat from the Persian Empire, which had extended westwards into Anatolia, and was now approaching the shores of the Aegean. In 494 BC the Persian army overran Miletus. The fate of Anaximenes remains unknown, and the Milesian school came to an end less than a century after it had begun. The ruins of Miletus remain to this day, now several miles inland from the silted estuary of the Meander.

But philosophy survived the fall of Miletus. By a combination of flukes and legendary heroism, the Greek world managed to resist, and even overcome, the great invading Persian armies. One of the high points of this resistance was the battle of Marathon in 490 BC, where a heavily outnumbered Greek force put the entire Persian army to flight. (The messenger delivering news of the victory collapsed and died after running the twenty-six miles and 385 yards from Marathon to Athens. This event is commemorated in the modern marathon, which is run over the identical distance.)

By the fall of Miletus, philosophy had already spread from the Ionian coastline to the Aegean islands, and thence to the rest of the Greek world. Ephesus was the major city of Ionia, and survived the Persian onslaught by the simple expedient of allying with the enemy against its commercial rivals such as Miletus. Its best-known philosopher, Heraclitus, was similarly problematic. Heraclitus was born around 540 BC, and was an arrogant, misanthropic man. In old age he became so disgusted with his fellow Ephesians that he forsook the city for a wandering life in the mountains, living off grass and herbs. His philosophy on the other hand was calm, subtle and profound. Heraclitus had his own idea about the fundamental element from which the world was formed. According to him, this was fire.

Anaximenes had understood the need to explain the diversity of the world. Heraclitus saw that Anaximenes had only partly answered this question. What did it mean to say that air turned into water, earth, stone and so forth? If it didn't remain air, this couldn't be the one substance out of which the world was made. Here was a complex and serious question. Heraclitus recognized that it was unanswerable as long as the 'one' was seen as a substance – such as air, or even water. It was necessary to view the 'one' as immaterial. 'The world has been, is now, and ever shall be an ever-living Fire, kindling by degree and going out by degree.' For Heraclitus the world was in a continuous state of flux. This is best understood in his celebrated sayings: 'No man steps into the same river twice' and 'The sun is new every day.' Fire was conceived as the underlying pattern or order of the cosmos, transforming itself yet remaining the same.

Scientifically minded philosophers were for centuries inclined to dismiss Heraclitus' notions as mere mysticism – until the advent of twentieth-century science. Only then did the scientific subtlety of his thinking become apparent. Heraclitus' ever-changing fire uncannily resembles the notion of energy in modern physics. Here was a philosophical outlook which could accommodate relativity and the ambiguities of quantum physics. (In relativity mass is equivalent to energy, according to Einstein's $E = mc^2$. Thus energy can theoretically transform itself into matter, just like Heraclitus' flux or fire.)

Heraclitus was to meet a sorry, self-inflicted end. His meagre diet of grass and herbs finally forced him down from his mountain solitude back to Ephesus. However, far from being skeletal he was by now bloated with dropsy, which causes the sufferer to retain fluid in the tissues. Displaying characteristic arrogance, he demanded a cure from the local physicians in the form of a riddle: 'Can you create a drought after heavy rain?' When the physicians proved baffled by this meteorological test, he decided to cure himself. Retiring to a cowshed, he buried himself in manure, apparently in the hope that the noxious fluid in his body would be drawn out by the warmth of its noxious coating. This drastic method proved ineffective, and he died a noxious death.

The ultimate element had now been seen as water, as air and as fire. So which was it? There seemed an obvious answer to this question. Why should there be just one? Why not several? Why not all three of them – with an added fourth element to account for the world's solidity? The clear answer seemed to be that the world was in fact made of four fundamental elements: earth, air, fire and water.

This notion of plurality liberated at a stroke scientific thinking from the stranglehold of unity. It was the obvious compromise – and with hindsight we can see that it pointed the way to a dramatic new understanding of the elements. Sadly, this can only be seen in hindsight. This 'obvious' compromise – the notion of four basic elements – was to prove one of the biggest blunders in human thought, and its effects were to be a catastrophe for our intellectual development.

When early philosophy spread through Ionia and the ancient Greek world, it was like a sudden clarification of the human mind. Evolution was focusing the lens of human thought: what had previously been a blur of superstition and metaphysics was now becoming clear. We could see! Human beings were learning how to look at the world around them. Conversely, the notion that the world consisted of four elements was like a disease, and it was to cripple scientific thinking for the next two millennia.

This was by no means the fault of the ingenious philosopher who first thought up this idea. The four-element theory was put forward by Empedocles, who lived in a Greek colony in Sicily during the fifth century BC. He was influenced by Pythagoras, who remains one of the great enigmatic figures of the early Hellenic era. Pythagoras ranks amongst the finest mathematicians of all time. He discovered that π was incommensurable and proved the theorem named after him. His other role as the founder of a religion based on a mixture of pure spiritual insight and pure hocus-pocus also betokens exceptional talents. Empedocles was to follow in the footsteps of his master as a thinker and charlatan. His theory of the four elements was a stroke of genius, but he didn't just leave it at that. He extended this scientific world-view with further theories of the highest order. He claimed that nothing in the world was created or destroyed – and maintained that all things consisted of differing combinations of the four elements. Here, for the first time, are the shadowy beginnings of the idea of chemistry.

Again like Pythagoras, Empedocles was that curious combination of a man ahead of his time and a man behind his time. This schizophrenic straddling of two ages has produced some of the most original minds in history. (One has only to think of Shakespeare's combination of the medieval and the Renaissance mind, of Newton's contradictory devotion to alchemy and mathematical physics.) Empedocles was an early version of this type. Although most Greek philosophy was by now being written in prose, Empedocles chose to hark back to a previous era by confining his scientific thinking within the straitjacket of poetry. His finest work, *On Nature*, was a five-thousand-line epic, of which we know only fragments. This masterpiece seemingly consisted of a heady cocktail of brilliant original ideas and downright quackery. Amongst the former was the idea of evolution. No mere flight of poetic fancy, this was a fully-fledged theory worked out in some detail. Empedocles envisaged evolution in terms of anatomical units: limbs, organs, heads and so forth. These combined in different ways. To begin with there were all kinds of odd combinational creatures – such as 'man-faced ox-progeny' (just like the centaurs and satyrs of Greek mythology). Only the creatures which were best adapted to their surroundings survived. But as ultimately nothing was created or destroyed – in other words the ingredients of the world remained the same – he was able to claim: 'I have already been once a boy and a girl, a bush and a bird and a leaping, travelling fish.' It was to be over two thousand years before scientific ideas of a similar calibre about evolution resurfaced in Western Europe.

The division between genius and nonsense is sometimes wafer thin – a barrier easily transcended by the truly convincing charlatan, who on occasion even succeeds in convincing himself. Empedocles was to die when he leapt into the crater of Mount Etna, in an attempt to prove to his followers that he was immortal. Opinion remained divided at the time, but over the years his lack of reappearance went against him.

Empedocles' notion of the four elements may have been wide of the mark in theory, but ironically it showed considerable insight into the practical side of chemistry. Earth, water, air and fire take on a much more significant aspect if we look at what exactly they are. Earth is a solid, water a liquid, air a gas, and fire could easily be seen as energy. Here is an eminently practical division of substances into different types – a classification one might expect of an embryo practical chemist rather than a theoretical philosopher.

Yet this practical notion of elemental classification is not really so surprising. The idea of chemistry was at this stage no more than a few shadowy insights, but the unwitting practice of it was already well under way. The ancients had long known of chemical processes. The earliest recognizable chemists were women, the perfume-makers of Babylon, who used the earliest known stills to produce their wares. The first individual chemist known to history was 'Tapputi, the perfume-maker', who was mentioned on a cuneiform tablet from the second millennium BC in Mesopotamia.

Practice long preceded theory. By the Hellenic era the ancient world had also discovered over half a dozen metallic elements, and a couple of non-metallic ones. The ancient Egyptians knew gold and silver, copper and iron; all four are also mentioned in the Old Testament. The Phoenicians are known to have used lead to weight their wooden anchors, though when they sailed to Spain they found more silver than they could transport, so they discarded their lead and weighted their anchors with silver instead. Subsequently they travelled even further afield, to Britain, where they traded in another metallic element from the Cornish mines – namely, tin.

Indeed, it was bronze, an alloy of tin and copper, which gave its name to the age which began in the Mediterranean region around 3000 BC. It was in the Bronze Age, around 1250 BC, that the Mycenaeans overran Troy. Hard metallic bronze is formed when tin and copper are melted together. This alloy was used for ornaments and tableware. But its most significant use was in weapons and armour. Some two millennia later bronze was superseded by a harder alloy made from smelted iron and carbon, instigating the Iron Age.

The other metallic element known to the ancients was mercury, which is mentioned in ancient Chinese and Hindu texts, and has been found in Egyptian tombs dating from 1500 BC. Owing to its exceptional appearance and qualities (mirror-surfaced, liquid metal, highly poisonous) mercury was regarded with awe, and considered magical from the outset.

As for the non-metallic elements, carbon and sulphur were certainly known from earliest times. The Roman author Pliny refers to long-established Sicilian sulphur mines whose product was used for medicinal purposes and for sulphur matches. Carbon was known to cavemen in the form of soot and charcoal. In its hardest and most precious form, diamond, this element was mentioned in the Old Testament and the Hindu Vedas, in texts dating from the second millennium BC.

Both the ancient Greeks and the Romans knew of a substance which they called 'arsenic'. But this was not the pure element – it was the sulphide; they used it for curing hides and poisoning rivals. And here lies the key. The ancients knew of these elements, but they didn't know them as such. They had no idea that they *were* elements. Such a thing remained beyond their conception. The notion of an element originated with the philosophers, not the chemists. In other words, with the thinkers, not the practitioners. Thales' theory that the ultimate element was water was the true beginning: the scientific idea of what an element is.

Anaximenes developed this, and a short while afterwards another startlingly original scientific idea was conceived and developed by the ancient Greeks. It was the fifth-century philosopher Leucippus who asked the question: 'Is matter discrete or continuous?' In other words, is it possible to go on dividing things up indefinitely, or does one reach a point where things become indivisible? Leucippus considered it self-evident that the latter was the case. This led him to the idea of the *atomos*, or atom. In Greek this word means 'uncuttable', i.e. indivisible. Leucippus was the first to state that the world was made up of indivisible atoms. Astonishingly, he came to this conclusion only a century after Thales had initiated scientific thought. Leucippus was probably born in Miletus, but must have left before the Persian invasion. It seems he set up a school at Abdera, on the northern Greek mainland. Here his best-known pupil was Democritus, who was to develop Leucippus' original atomic idea. According to Democritus, there is an infinite amount of atoms, which exist in perpetual motion in space; there are also innumerable different kinds of atoms, which differ in shape and size, weight and heat. All apparent change in the world, he argued, is due to combinations and recombinations of these unchanging atoms.

Much like the ideas of elements and of evolution, Democritus' ideas appear breathtakingly modern – far, far ahead of their time. Such originality remains unprecedented in human thought. Once the Greeks had come up with philosophic thinking, it appears they quickly thought it through almost to its limits. No less a figure than Bertrand Russell claimed: 'Almost all the hypotheses that have dominated modern philosophy were first thought of by the Greeks.' But it was to be a double-edged legacy. As we shall see, when Greek thinking took a wrong turn it could put human intellectual development on the wrong course for centuries to come.

During the lifetime of Democritus Greek philosophy entered

its golden era, in Athens. This had become the richest and most powerful city in the Greek world, the apex of Greek culture. Its acropolis was crowned by the sublimely proportioned pillars of the Parthenon, one of the finest architectural achievements in history. In the city's panoramic, acoustically superb, open-air theatres the tragedies of Aeschylus, Sophocles and Euripides were performed. (It was here that Greek tragedy developed from a primitive religious ritual to profound drama on a proscenium stage in the course of a single generation.) But by the time philosophy entered its golden era, Athens had begun to decline, under the impact of the long and disastrous Peloponnesian Wars against Sparta. Ironically this golden era, which remains unrivalled in the entire history of philosophy, was to be very much a reaction against the achievements of Democritus, Empedocles, Heraclitus and their like.

The first great Athenian philosopher was Socrates, who was born around 477 BC and was sentenced to death in 399 BC. Socrates was one of the great characters of philosophy, by turn endearing and infuriating. Pronounced 'the wisest among men' by the Delphic oracle, he claimed he knew nothing – and then proceeded to demonstrate that others knew even less. Before his young pupils in the agora (market place) he would sometimes use this 'dialectical method' to criticize the learned worthies of Athens – a course of action which won him few friends in high places, and almost certainly played a part in his later trial and death sentence for 'corrupting youth'. Socrates' dialectical method consisted of pretending to know nothing, and then questioning the knowledge put forward by his adversary, asking its meaning, breaking down the concepts upon which it was based. This was an early form of analysis. (The Greek word *analytika* means 'to unravel'.)

Analysis is always necessary to clarify meaning – but the way Socrates used it, this method tended to take knowledge apart

rather than build it up. An embryo science, whose conce are vague and whose knowledge is highly theoretical, can withstand this kind of attack, and Socrates found it easy to pi holes in such 'learning'. (To this day science betrays simil inadequacies; its redeeming merit is that it works in reality, rathe than in philosophical terms.)

Socrates was not interested in atoms or evolution or the fundamental elements which make up the world. Instead, his philosophy was based on introspection – his favourite saying was 'know thyself'. Here was the only true knowledge. With disastrous consequences, philosophy turned away from the world.

Socrates was succeeded by his great pupil Plato, who persevered with this tradition. Instead of scientific speculation, philosophy now turned its attention to abstract ideas. Mathematics, truth and beauty were prized far above 'mere appearances'. The particularities of the world around us were just an accidental jumble of chimera – it was only ideas that were real.

Plato was born into an aristocratic Athenian family, and was to become the supreme philosopher of the classical era. It has been said that he formulated every major question in philosophy, and that all philosophy since has consisted of little more than footnotes to Plato. There is some truth in this, but it is far from being the whole truth. It certainly does not apply to so-called 'natural philosophy', i.e. science. Plato could not have asked 'What is electrical force?' because he had no understanding of what electricity is. This may seem obvious, but it is crucial. Although science is no longer regarded as part of philosophy, it certainly has its effect on our understanding of knowledge and ourselves. As a result of Copernicus, Darwin and Freud, our idea of ourselves is fundamentally different from that of Plato and his contemporaries. We can understand Greek tragedy, and empathize with the stark force of its emotions, but we neither think nor behave in this fashion any more.

In 387 BC Plato opened his Academy in an olive grove on the outskirts of Athens. This original 'grove of academe' was the first recognizable university. Above its entrance was written: 'Let no man enter here who does not know geometry.' Abstract reasoning, abstract ideas, abstract geometry – even Plato's political teaching concentrated on the idea of a utopia rather than on social reality. The Academy was supreme in mathematics, but the geometry taught here was limited to figures which could be drawn with a ruler and compass. Only such figures were ideal (i.e. corresponded to 'real' ideas); all others belonged to the accidental jumble of the actual world (which was simply an illusion, 'mere appearance'). The latter were dismissed as 'mechanical'. Geometry was concerned only with such as the circle, the triangle and the regular polygon – none of which appears precisely in nature. Democritus had arrived at the concept of the atom by applying mathematics to nature. (Apply division to the world until you can go no further.) Plato was only interested in applying mathematics to itself. (This attitude lingers on in the notion of 'pure' mathematics.)

The third of the Greek triumvirate of great philosophers was Aristotle, who was a pupil of Plato. Where Socrates had been 'the gadfly of Athens', and Plato the philosophers' philosopher, Aristotle was the first universal genius. He travelled throughout the Aegean, and at one point became tutor to the young Alexander the Great – though no record remains of what the ancient world's greatest polymath taught (or tried to teach) the ancient world's greatest megalomaniac. Aristotle was to make major contributions in almost every field except mathematics. He was interested in everything, and his library reflected this. Never before had such a collection of scrolls been accumulated by a private citizen. With Aristotle the anti-scientific bias came to an end. But by now the damage had been done. Although Aristotle was one of the finest scientific minds in all history, his ideas were fatally infected with Platonic thinking. (Which still echoes in our use of the word: not

The classic bust of Aristotle. As this is a Roman copy of a contemporary Greek bust, it probably bears a strong resemblance to Aristotle's actual appearance

for nothing is a platonic love affair one in which nothing actually happens.) Aristotle's achievements mapped out the course of scientific development until well into the modern era. He made major contributions in every field of natural philosophy from botany to geology, and psychology to zoology. Indeed, it was he who delineated many of these scientific fields in the first place. And his supreme achievement was the invention of logic.

So where did he go wrong? Aristotle turned Plato's philosophy on its head, believing that ideas could exist only in the particular substance that embodied them. Nonetheless his thought remained tainted with the Platonic way of seeing the world. He saw objects as possessed of qualities – the Platonic ideas which inhabited them – rather than actual properties. There is a subtle but fundamental difference here. A world consisting of substance or objects may have *qualities*; a world composed of atoms has *properties*. You don't see ideas embodied in an atom. Only in this way could a man of Aristotle's brilliance have accepted earth, air, fire and water as the four basic elements. These are qualities.

When Aristotle applied the brilliance of his scientific mind to this error, the error was compounded with disastrous results. Using reason and observation, Aristotle worked out that each of the four elements had its place. Earth was in the centre, next came water, above that was air and above the air came fire. All motion in the world was the attempt of the elements to find their rightful place. Thus stones fell to the bottom of water, bubbles rose up out of it, fire rose upwards through air and so forth.

But the sun, the moon and the stars clearly did not move in such a way, so Aristotle proposed a fifth element. This rarefied element he called aether. Remnants of this element persist in our words ethereal (celestial, airy) and quintessence (literally 'fifth essence'.) The heavens, from the moon upwards, were all part of this aethereal realm. Being nothing to do with the other four elements, the heavenly bodies were not subject to the same laws, which explained why they did not fall to earth. This upper aethereal realm contained concentric transparent spheres of crystal, arranged about the central earth like the layers of an onion. The moon, the sun, the planets and the stars were all embedded in these spheres, whose separate rotations produced the movements of the heavenly bodies. As the transparent crystal spheres moved against each other they generated sublime harmonies which were inaudible to the human ear, the 'music of the spheres'.

Such was Aristotle's authority, and the surpassing brilliance of his thinking, that all this was accepted. The earth was taken as the centre of the universe, despite several pre-Socratic thinkers having realized otherwise. And this, coupled with the notion that the heavens consisted of a different element and obeyed different laws, effectively finished off astronomy until the advent of Copernicus in the sixteenth century.

Aristotle, like Empedocles and Shakespeare, was a genius who straddled two eras. His historical position is perhaps best illustrated by his attitude to politics. Not for Aristotle the impractical utopia of Plato – this was a subject to be considered scientifically. In order to discover the best political system, Aristotle collected in his library scrolls containing the constitutions of all the city states in Greece. If any city state was to approach him for constitutional advice, Aristotle could draw up a constitution which not only fitted the circumstances of that particular city, but also contained the best points from the constitutions of all the other city states. If anything worked, this surely would. But it didn't, and it couldn't.

Why not? Ironically, the blame for this lay with Aristotle's pupil, Alexander the Great. It was Alexander who succeeded in uniting the Greeks, a feat he achieved by the simple expedient of conquering them. He then decided that his united Greece should no longer face the threat of an invading empire – so he set off east to conquer the Persians. Having done this, he determined to press on and conquer the entire known world. (Leaving aside the question of megalomania, it could be argued that Alexander's strategy was misguided. By this time the Persian Empire was in decline, while to the west the Roman Republic was just beginning to expand. A century and a half after Alexander's glorious campaign, which conquered all to the east, the remnants of his empire would begin to fall to invasion from the west.) Despite its shortcomings, Alexander briefly established the largest empire the world had yet seen. The city state (or *polis*) was replaced by the metropolis (literally 'parent city'.) The era of city states was over, the era of empires had begun. Democracy had been replaced by imperialism. For all Aristotle's intellectual brilliance, his thinking on how to achieve the best political constitution had been rendered utterly redundant.

If only the same could have happened to his elemental scientific thinking. The idea that the world consisted of earth, air, fire and water belonged in the past. And no amount of brilliant thinking could save any science based upon such a premise. The investigation of the elements would now be approached from an entirely different direction – one which was both unscientific and irrational. The science of the elements now passed into a darker realm.

2

The Practice of Alchemy

Alchemy is traditionally said to have begun in Alexandria. This city was founded at the mouths of the Nile in 331 BC by Alexander the Great, as his capital of the conquered territories of Egypt. Within two centuries it had become the largest city in the world, a thriving melting pot of Egyptian, Greek and other Levantine cultures. Its harbour contained one of the seven wonders of the world, the Pharos. This 450 foot high lighthouse had a beam, concentrated by a reflector, which could be seen out at sea from beyond the horizon.

But the pride of Alexandria was its Temple of the Muses (or Museum), whose library became the finest of the classical era. This contained over seventy thousand books (in the form of scrolls and papyruses), and attracted scholars from all over the Mediterranean world. The library established Alexandria as arguably the greatest centre of learning in ancient times, superseding even the Athens of Plato and Aristotle. The major intellectual figures of the Hellenistic era, such as Euclid, Archimedes and Aristarchus (the Copernicus of antiquity) studied there – as, later, did the astronomer Ptolemy and the first great woman mathematician-philosopher, Hypatia. Aristotle's extensive private library was eventually incorporated into it, and his philosophy played a central role in the teaching faculty which became attached to it.

But in Alexandria Greek thought encountered a much older form of learning, known as the Egyptian art, or *khemeia* (the root of our word chemistry.) The origins of *khemeia* are lost in time. The word occurs in various ancient Egyptian hieroglyphs in connection with the burial of the dead. The Roman historian Pliny even says that Egypt itself was originally called *khemeia*, or black, after the dark rich soil of the Nile delta. The knowledge associated with this dark art is mentioned in several early sources, including the Book of Enoch (one of the apocryphal books of the Old Testament). According to this source the secret knowledge of *khemeia* was passed on to a number of women by fallen angels in order to win their favours.

Initially this knowledge consisted largely of the chemical processes involved in embalming the dead. This was necessary, to preserve the corpse for its journey to the world of the dead. Owing to this association with the underworld, the practitioners of *khemeia* came to be regarded as magicians and sorcerers. However, its practices soon evolved to include other chemical processes which had been discovered by the ancient Egyptians, such as glass-making, dyeing and, especially, the art of metallurgy. *Khemeia* thus became connected with the seven known metallic elements: gold, silver, copper, iron, tin, lead and mercury. This, and its association with the dead, became the foundation for a body of metaphysical knowledge. The Egyptians noticed that, like the seven elements, there were also seven planets, or 'wandering stars', which moved against the background of the fixed stars. These were the Sun, the Moon, Venus, Mars, Saturn, Jupiter and Mercury. It was not long before they made a connection between these two groups of seven. The Sun became associated with gold, the Moon with silver, Venus with copper and so forth. (Our modern name for the metal mercury comes from that of the planet.) As with astrology, early alchemists suspected their science had uncovered one of the secrets of the universe. This showed how the earth (and hence humanity as well) was related to the cosmos.

In purely practical terms this connection also gave the wizards and magicians a method of protecting the secrets of their trade. Instead of describing how to alloy two metals to form bronze, they could refer to a conjunction between Venus (copper) and Jupiter (tin).

They could also describe various improving spiritual practices in terms of transmuting base elements (primitive behaviour) into gold (nobility). All this knowledge was attributed to the Egyptian god of wisdom, the ibis-headed Thoth. When the Greeks encountered this god they identified him with their own god Hermes, the messenger of the gods. Hence the occult practices of alchemy became known as the 'hermetic art'.

But it was when the Greek philosophical tradition encountered *khemeia* that alchemy was truly born. The philosophers had succeeded in separating science from religion. Now, in a retrograde step, the two were to be reunited. And worse still, this error was compounded: the practitioners of *khemeia* knew of seven genuine elements, but these were discarded in favour of the four Aristotelian elements.

What at first appears to us as a disaster, however, became a new source of inspiration. Once *khemeia* adopted earth, air, fire and water as the four elements, it was soon recognized that these were not fixed elements. They were qualities – analogous to hot and cold, wet and dry. But qualities could be changed. Hot could be turned into cold, wet into dry and so forth. Yet if elements could be changed, perhaps there was some literal truth in those mysterious ancient texts which appeared to allude to the transformation of base metal into gold. (With the benefit of hindsight, we can see that this introduces yet another error: the muddling of physical change with chemical change.)

A central aspiration of alchemy now emerged. Neither spiritual wisdom nor chemical technique was sought – the aim was pure gold. There appeared to be no hope for science here.

Yet, ironically, an embryo science did begin to emerge. As the alchemists pitted their wits against the elements, they were to come up with a body of scientific knowledge which complemented that of the philosophers. The Greek philosophical tradition was now becoming moribund as bit by bit the Roman Empire took over the remnants of Alexander the Great's old empire. The Romans were practical, rather than thinkers, and added nothing original to Greek thought. (It is said that the only Roman who appears in the history of mathematics is the one who slew Archimedes.)

Alchemy now mirrored this decline. It lapsed into semi-mystical mumbo-jumbo; its creative scientific energies were invested elsewhere, in the practical pursuit of gold. Many and ingenious were the methods devised to crack this particular nut. Intellectual endeavour (or at least this branch of it) may have lowered its sights, but low cunning now came into its own. The grubby wizard-charlatan in his soot-blackened den took over from the toga-clad philosopher expounding beneath clear skies. In their day both were equally the subject of popular ridicule, the invariable fate of originality. Despite this, both alchemy and philosophy were to survive after their own fashion. Yet it is arguable that we owe more to the former than to the latter. Today we can, at a pinch, do without philosophy, but we can't do without chemistry.

The earliest known adept of the dark arts was a certain Bolos of Mendes, a Hellenized Egyptian who lived around 200 BC. In his principal work, later called *Physica et Mystica* ('The Physical and the Mystical'), he listed a wide variety of esoteric experiments, each of which ended with the incantation: 'One nature rejoices in another. One nature destroys another. One nature masters another.' This chant contains a tantalizing hint of true chemical understanding. (Could this be a description of how some substances dissolve in others, some corrode each other and some form compounds?) Unfortunately the actual experiments Bolos describes would seem to belie such understanding, being devoted mainly to various methods for producing silver and gold. These were heavily larded with references to Greek philosophy and cosmology, and remained obscure enough to withstand proof or disproof through the ages.

But the truth will out. In 1828 an ancient papyrus was discovered in Thebes. This listed experiments remarkably similar to those given in *Physica et Mystica*, but stripped of all theoretical and decorative obscurity. The papyrus made it plain that the methods it described for producing gold and silver were fraudulent, intended only to dupe the innocent. Unfortunately the intervening two millennia had seen the consequent rise, flourishing and eventual decline of alchemy, unaffected by this genuinely hidden piece of knowledge. So Bolus was a fake, but other indications (not least his fakery) seem to suggest that he had an inkling of the fundamental nature of chemistry, i.e. that the elements themselves cannot be changed by chemical experiment.

The same cannot be said of Zosimos of Panopolis, acknowledged as the greatest of the early alchemists, who practised in Alexandria around AD 300. (Alchemy remains a 'practice' in the musical rather than the medical or scientific sense: this pre-performance stage was not to be succeeded by any actual performance which produced the goods.)

Zosimos compiled an encyclopedia of alchemy in twenty-eight volumes, one for each letter of the Greek alphabet. (The ancient Greek alphabet in fact had only twenty-four letters, but temporarily acquired more during the Byzantine era.) The contents of this encyclopedia bear a similarly loose attachment to reality, being written in a style by turns cryptic, mystical and nebulous. Symbolic formulae and coded experimental instructions abound. Just occasionally the clouds of mysticism and metaphor part to reveal a glimpse of something less vaporous. At one point Zosimos defines alchemy as the study of 'the composition of waters, movement, growth, embodying and disembodying, drawing the spirits from bodies and bonding the spirits within bodies'. Here mysticism and genuine chemical practice appear to coexist, metaphorically. But which is he really talking about? The ambiguity is typical. Other passages, such as those dealing with the 'ennobling' of base metals into gold, are obscure for more obvious reasons. Yet even these contain tantalizing glimpses which suggest a perceptive understanding of chemical practice.

Experiments are described which involve several distinct stages in a chemical process.

> Mix the yolks of eggs with the grindings of their shells. Pour the mixture into a sealed container and burn for forty-one days. Then leave the container to cool on the embers of a sawdust fire. You will now find its contents transformed into a completely green substance. Boil this residue with water, and the solution will vaporize to become divine water. Do not touch this with your hand, only with an instrument made of glass. Put the divine water in a sealed container and cook for two days. Then empty the contents on to a conch shell, smooth them out and expose to the sun. The water thickens into a soapy substance. Melt an ounce of silver, add this substance, and you will have gold.

The stages of an experiment are frequently characterized by a particular colour – evident in the residue, vapours or solution – marking the successful completion of a particular treatment. (In one case, for example, black, white and green stages must be passed through before the red which heralds the appearance of gold.) In many of these experiments coloured compounds – sulphates and sulphides in particular – are distinctly recognizable. Experimental procedures such as distillation, filtration and solution are also recognizably described. One ambiguously worded passage even suggests that Zosimos may have been the first to isolate the element arsenic. But it seems more likely that he was referring here to a compound, probably arsenic sulphide. Perhaps most interesting of all, Zosimos appears to have understood that chemical change can be induced by the presence of a catalyst. This is an extra substance introduced into an experiment which causes the other ingredients to react, or speeds up their reaction, but which itself remains unchanged at the end of the process. Zosimos describes several chemical processes for turning base metals into gold that involve a catalyst, which he refers to as a 'tincture'.

Yet the obsession with transmuting base metals into gold did not entirely monopolize alchemical practice. Ironically, mystification could sometimes result in a scientific bonus – two wrongs making a right, so to speak. Some of the alchemical procedures described by Zosimus speak of 'curing sick metal' by turning it into gold. Not surprisingly, the occasional tyro alchemist misinterpreted this, and set off on the wrong track altogether. In this way, alchemy inadvertently became useful. Instead of curing sick metals, these misguided alchemists sought ways to cure actual ailments. Having laid some haphazard foundations for the science of chemistry, alchemists now accidentally founded scientific pharmacy, the chemical approach to medicine.

But this was far too good to last, and alchemy soon progressed from curing mortal ailments to curing spiritual ones. As in much improving literature, the subtext of allegory, symbols and metaphor was far too strong for the simple storyline of interacting agents, human chemistry and so on. From avarice, to medicine, to salvation: thesis, antithesis, synthesis – alchemy's evolution followed its wayward dialectic course. And then, being only human, returned to where it had started from: gold!

This trio of avarice, medicine and salvation was to remain an integral part of alchemy throughout its development. Indeed, they played a central role not only in Alexandrian alchemy, but also in the alchemies which developed in other parts of the globe around the same period. By the early centuries AD recognizable alchemical practices were well established in South and Central America, China and India. All of these practices shared the same trio of mixed motives, and these separate regions developed their alchemy independently of Alexandrian alchemy. (The possible exception here is India, which may have received influences from the West via Gandhara, the Greek state set up in India by Alexander the Great, which persisted for several centuries.) Such independent development suggests that alchemy formed a universal stage of human evolution, a necessary stepping-stone in our intellectual development. There were of course differences too, characteristic of the societies in which they evolved. Western alchemy, for instance, always remained obsessed with making gold. Chinese alchemy, on the other hand, concentrated more on medicine and salvation, developing a blend of these two aspects in the search for immortality. Hence the Chinese 'elixir of life', which promised either eternal youth or immortality. (The word elixir is not Chinese, and in fact dates from many centuries later – but this is the word which best evokes the character of this substance.) These elixirs were taken by those who enjoyed life so much that they wished to go on living for ever. In China, only the ruling caste fitted into this category. Unfortunately, as the elixirs became more sophisticated, they also became more poisonous. According to the historian of science Joseph Needham, it now looks as if a whole series of Chinese emperors died of elixir poisoning. The elixir of life soon appeared in the west, independently it would seem, but inspired by similar credulous motives.

It is easy for us to mock at the aims which inspired alchemy in its development: untold riches, panaceas, immortality, etc. Yet this is the practice which was to give us chemistry. And what have been the aims of modern chemistry? Our investigation of the elements (and, ironically, the successful transmutation of elements, by splitting atoms) brought us to the brink of self-destruction. Beside this, the avarice, confidence tricks and charlatanry of early alchemy appear as harmless foibles.

But the days of Alexandrian alchemy were numbered. The secret ways and incomprehensible writings of its wizards and sorcerers meant that it showed no sign of becoming integrated into society, either as an intellectual practice (like mathematics) or as a religion (like the growing cult of Christianity). In AD 296, just a few years after the death of Zosimos, the Emperor Diocletian banned alchemy throughout the Roman Empire, ordering all alchemical writings to be burnt. The destruction was wholesale; this is one of the reasons why our knowledge of early alchemy remains so sketchy. Diocletian's edict contains the first official mention of the word *khemeia*. As with many of the complex allegories of alchemical writing, the first was intended to be the last. (Paradoxically, Diocletian had banned alchemy because he thought it would be successful. He was afraid that the widespread production of gold would undermine the empire's tottering economy.)

A century later, in AD 391, the Alexandria Library was sacked and burned to the ground by Christians. A vast compendium of classical learning went up in flames, vanishing for ever from the face of the earth, with only stray references in other works to indicate the huge scale of the loss to humanity. The practice of *khemeia* would have had no place in the library, but its more scientific findings might well have been noted by a few open-minded natural philosophers. Was some alchemist actually the first to isolate arsenic? Did some savant of *khemeia* return to the atomic thinking of Democritus, abandoning the error of Aristotle's four elements? If so, important discoveries would almost certainly have resulted – but we will never know.

Alchemy now disappeared underground, beyond the reach of history. Curiously it was a Christian sect, the Nestorians, which seems to have carried the secrets of alchemy to its next destination. By now Christianity had been declared the official religion of the Roman Empire by the Serbian-born emperor Constantine. In a move to unite the strife-torn empire, Constantine moved his capital east to Byzantium (the site of modern Istanbul) on the shores of the Bosphorus.

This city would soon become known as Constantinople, after the emperor. Within decades its nobility had become renowned for their extravagance. Many owned over a dozen houses, with a thousand slaves at their disposal. In their palaces ivory doors parted to reveal mosaic-floored halls, where couches plated with gold and encrusted with precious stones were laid out beneath silk hangings and censers spilling incense. (Taste was never a strong point with the Byzantines.)

In 325 Constantine summoned the Christian leaders from all over the empire to a council at Nicaea (modern Iznik), on the shores of the Sea of Marmara. These leaders believed in widely differing interpretations of the Scriptures, but Constantine browbeat them into accepting an official line on such matters as the divinity of Christ, and his equality with God. This was essentially a political move, enabling Constantine to strengthen his hold on the empire by merging the power of the Church with that of the state. Nevertheless heresies continued to plague the empire. Just over a century later, in 431, the Council of Ephesus was called to deal with the problem of Nestorianism. This Christian sect believed that Christ had two separate natures – divine and human – and was in fact two people in one. Nestorianism was now declared a heresy and its followers were forced to flee east to Persia. Here their schizoid heresy was eventually tolerated by the local Zoroastrians, who worshipped fire.

A number of Nestorians had continued with the clandestine

practice of alchemy, and they took with them the secrets of the dark art. These eventually passed on to the Zoroastrians, who were intrigued by the mysterious manifestations of fire in alchemical theory and practice.

In Asia, this esoteric branch of European learning began to flourish. Meanwhile, other more plausible branches of European learning withered. In 529 the Christian Emperor Justinian closed down the nine-centuries-old Academy of Plato in Athens, declaring it to be a centre of 'pagan learning'. This date is traditionally recognized as the beginning of the period once evocatively known as the Dark Ages, which were to envelop Europe for the next five centuries.

By now the barbarians had overrun the western half of the divided Roman Empire. Only the eastern Byzantine Empire remained, with its capital Constantinople at the eastern edge of Europe, facing across the Bosphorus to the shores of Asia. The ancient world had produced Archimedes, supreme mathematicians, and extraordinary feats of Roman engineering. Now, for almost seven centuries, Europe would not produce an original scientist worthy of the name.

The first great external threat to the Byzantine Empire emerged from the east. (Paradoxically, this threat would prove to be the saviour of European science.) Previously the Arabs had been insignificant desert nomads occupying the arid regions of the Arabian peninsula. In the seventh century all this was to change, under the inspiration of one man – a trader from the oasis of Mecca who in 610 had a vision of the archangel Gabriel. In this vision, the forty-year-old trader Muhammad was ordered to lead his Arab people on a mission which would bring the religious tradition of Judaism and Christianity to its ultimate fruition. This could be done only by submission to the one true God – Allah. Thus was founded the religion of Islam. (In Arabic the word *islam* means 'humility or submission'; 'salaam' and 'Muslim' derive from the same root.)

Muhammad succeeded in uniting the tribes of the Arabian peninsula under Islam. Fired with religious zeal, the Arabs now embarked upon a campaign of conquest the likes of which had not been seen since Alexander the Great. Within little over a century the Arabs would have an empire stretching from Spain in the west, across all of north Africa and the Middle East, as far east as northern India and the frontier of China.

In 670 the Arab fleet even laid siege to Constantinople, the last remnant of the Roman Empire and the centre of Christendom. Surrender was only a matter of time. The Arabs could then strike across southern Europe to meet up in a pincer movement with their compatriots who were advancing to Spain, and would eventually spill over the Pyrenees as far as Tours in central France. Europe was set to become a Muslim continent.

This plan was thwarted by an alchemist called Callinicus. Probably born in Egypt of Greek parentage, Callinicus had fled before the advancing Arabs, carrying with him the secret formula for 'Greek fire'. The secret of Greek fire is now lost, but its main ingredient appears to have been distilled crude oil. (Naturally occurring lakes of crude oil were common in the Middle East, and their flammability had played a significant role in inspiring the Persians to become fire worshippers.) The distilled crude oil produced a primitive form of petroleum, which was then probably mixed with potassium nitrate (as a source of combustible oxygen) and quicklime (which reacts with water to produce heat). When Greek fire was poured on the waters of the Bosphorus it set fire to the wooden hulls of the Arab fleet, and all attempts to douse it only caused further conflagration. The Arab fleet was destroyed, and Europe was preserved for Christianity, by this early 'secret weapon' – manufactured by the first and only scientist the Byzantine Empire was to produce.

The Arabs were suitably impressed by this example of Greek learning, and soon began discovering other examples of this ancient knowledge in Syria and Mesopotamia. From the Nestorians in Baghdad they learned the art of *khemeia*, which they soon called *al-chemia*. (The prefix *al* is Arabic for 'the'.) When the Arabs began to study and advance Greek learning, they started introducing their own words. Alcohol, alkali, algebra and algorithm are all Arabic in origin.

For the next five hundred years the history of chemistry, and that of most other sciences (including mathematics), was to remain almost entirely in Arab hands. The central core of Arabic alchemy appears to have been based on *The Emerald Tablet*, a work written by the legendary Hermes Trismegistos (Greek for 'Hermes the thrice-greatest'). This mysterious figure was variously said to have lived during the time of Moses (thirteenth century BC), to have been a descendant of the Greek god Hermes or his ancient Egyptian counterpart Thoth, and to have been the god Hermes himself. Through the ages his name had become attached to a body of works, the most important being *The Emerald Tablet*, which handed down many of the ancient secrets of *khemeia*, including formulae for the transmutation of base metals into gold.

The Arabs quickly saw great potential for this process, and set to work with a will. Greek learning in all its manifestations (from astronomy to philosophy, from mathematics to alchemy) now began attracting the finest Arab minds. The first exceptional figure to emerge in the field of alchemy was Jabir ibn-Hayyan, later known in Europe as 'Geber'. (For several centuries he was erroneously thought to have been the inventor of *al-geber*, or algebra.) Jabir was born around 760 and lived in Baghdad when the Arabic empire was at its height under the fabled Harun ar-Rashid (whose name prosaically translates as Aaron the Upright). This is the era evoked in the *Arabian Nights*, where Scheherazade eludes her sentence of death by telling stories to the caliph (supposedly Harun). Night after night she enthrals him with her thousand and one legendary tales of Aladdin, Sinbad the Sailor, Ali Baba and the like. Baghdad itself was if anything even more fabulous. By the ninth century this was the richest city in the world, its palm-fringed quays on the Euphrates lined with sailing ships from as far afield as Zanzibar and Cathay (China). At the city's heart lay the two-mile-wide Round City, guarded by three rings of walls. In the centre of the Round City stood the Golden Palace and the Grand Mosque, and from these four axial roads led out to the four corners of the Arabic Empire. Amidst the fountains and shady gardens in the suburbs outside the walls were centres of learning and public hospitals. By contrast the souks (vast covered bazaars, complete with balconies and turrets) were an exotic clamour of perfume sellers, sword-makers and armourers. Stalls offered cinnamon from Sumatra, cloves from Africa, even wondrous produce such as spinach and rhubarb (which were then unknown in Europe). In the busy plazas story-tellers recited tales of Aladdin and his magic lamp, fire-eaters and sword-swallowers competed for attention, and Indian silk merchants made bids for Nubian slaves. (Meanwhile in Europe, Rome lay in ruins; and in a draughty castle amidst the German forests Charlemagne gaped in awe at the intricate water-clock presented to him by a silk-robed Arab emissary.)

The arrival of alchemy in the Arab world was greeted with a similar intrigued astonishment. The first contact between the Arab mind and this remnant of Graeco-Egyptian thought produced a flurry of creative, if not always scientific, activity. The Koran positively encouraged medicine and the study of scientific and mathematical learning. This was the way to gain insight into the will of God. (Even in modern fundamentalist Islam this remains the case theoretically; it is for the most part the effects and products of such learning that are shunned.) Unhampered by such restraints, Jabir embarked enthusiastically on the quest for gold, and in doing so established himself as one of the greatest alchemists of all time. What is significant to us is that Jabir, unlike many of his more rapacious colleagues, approached the problem of transmutation in a scientific manner. His chemical thinking may have been based on false premises, but it remained analysis of the highest order. It was he who set chemists thinking of new possibilities in the ultimate construction of matter.

Jabir modified Aristotle's doctrine of the four elements, especially in regard to metals. According to Jabir, metals were formed out of two elements: sulphur and mercury. Sulphur ('the stone which burns') was characterized by the principle of combustibility. Mercury contained the idealized principle of metallic properties. When these two principles were combined in differing amounts, they formed different metals. Thus the base metal lead could be separated into mercury and sulphur, which if recombined in the right proportions could become gold. But like Zosimos, Jabir felt sure that this process required a catalyst, which would assist the process yet remain unchanged at the end of it. What Zosimos had called a tincture, other Greek sources such as Hermes Trismegistos referred to as *xieron* – meaning a dry or powdery substance. This became in Arabic *al-iksir* (elixir). Anything that could transmute base metals into gold surely had miraculous properties of its own. The elixir was soon seen as a medicine, capable of curing many diseases, then as a panacea (all diseases cured) and finally as the 'elixir of life', bestowing eternal youth or immortality. Entirely independently, Arabic thought had now reached the stage which Chinese alchemy had reached some eight hundred years previously. A reflection on the universality, to say nothing of the incorrigibility, of human aspiration.

Jabir's alchemical advances were significant in the embryo theory of chemistry, but he also made advances in what we would consider 'real' chemistry. The volatile and mysterious substance sal ammoniac (ammonium chloride), found around the lips of volcano craters, had been mentioned in *The Emerald Tablet*, but Jabir was the first to undertake a systematic investigation of its properties. This was a simple chemical compound which reacted with several commonly available substances to form other entirely different compounds. Jabir's investigations brought him to the brink of an understanding of chemical reaction – what precisely it is, and what takes place.

One of the commonly available substances which Jabir used was vinegar, which he distilled to form strong acetic acid. This acid had been known to the Greeks, who had been fascinated by its ability to dissolve certain substances. Jabir also managed to prepare weak solutions of nitric acid – which is potentially a much stronger acid. Here were being assembled the simple ingredients for any basic chemistry kit. The Arabs were approaching an understanding of chemistry and what it could do.

Jabir was able to practise his art owing to the protection of Ja'far, the caliph's vizier. Ja'far became Jabir's friend and patron, encouraging him in both his experimental work and in his writing. But life at the intrigue-ridden court of Harun ar-Rashid was a risky business. When Ja'far fell from favour, and was executed, Jabir was forced to flee to the safety of his native village. Here he lived out his days quietly, writing his masterpiece *The Sum of Perfection*, which includes a survey of all his vast alchemical and chemical knowledge.

Yet Arab alchemists remained principally interested in the pursuit of gold. This quest was to inspire the second great alchemist of the Arabic world, Al-Razi (later known in Europe as Rhazes). Al-Razi was in fact a Persian who flourished in Baghdad during the early decades of the tenth century. Like so many of the Arabic-world thinkers of the period, Al-Razi had a wide range of intellectual interests. He is known to have written an encyclopedia on music, insightful philosophy and poetry. His interest in science was prompted only by a casual acquaintance with a

Baghdad apothecary when he was in his mid-thirties. This sparked a consuming interest in medicine, which eventually led Al-Razi to be appointed as chief physician at the main hospital in Baghdad. His achievements in this field were both diagnostic and practical. Al-Razi's writings indicate that he was the first to recognize the categorical difference between smallpox and measles. These writings also describe how to prepare plaster of Paris, and how to use this to create casts for fractured limbs. In common with Jabir, Al-Razi's scientific writings are so clear and detailed that it is even possible to re-create his experiments in precise detail. Those who have done so attest to an exactitude and honesty in his findings which were not always the case during this period – except in mathematics, at which the Arabs excelled. Though, perhaps inevitably, Al-Razi's alchemical writings on transmutation do not depart from tradition. Here he prefers to retain the metaphorical impenetrability so necessary for the continuance of this pursuit.

Al-Razi's major work was called *The Secret of Secrets*. Fortunately this does not live up to the mystic promise of its title. It is in fact a clear outline of Al-Razi's chemical knowledge and practice, and as such one of the major scientific works of the Arab era. Al-Razi's forte, and much of his originality, lay in classification. At a certain stage every science requires a genius of classification, to enable it to be compartmentalized into various fields, so that these can advance in their own separate manner. The first great exemplar of this was of course Aristotle, who classified all the sciences known to the classical world. Al-Razi played a similar role for early chemistry.

The Secret of Secrets is divided into three parts. One part describes all the apparatus known to Arabic alchemy: a range of glassware and instruments which would be inherited by the chemistry laboratory and would remain largely standard until the nineteenth century. The next part describes 'recipes', i.e. the techniques known at that period, such as distillation, sublimation (solid to vapour), calcination (powdering of solids) and solution (dissolving of solids). Where the last two processes were concerned, the difference between physical change and chemical reaction remained hazy. By far the most interesting part of *The Secret of Secrets* is devoted to substances, and includes a long list of known chemicals and minerals. Here Al-Razi was the first to classify substances as either animal, vegetable or mineral. He also produces a list of the different types of materials used by alchemists. These include: 'bodies' (metals), stones, salts and 'spirits' (volatile liquids). In the last group he included mercury and sal ammoniac (ammonium chloride). Jabir's earlier investigation of sal ammoniac appears to have sparked a widespread interest amongst alchemists, who attempted to extend his experiments. In this way alchemists were now beginning to undertake geniune chemical research into the properties of various substances. And as we can see from Al-Razi's list of different types of material, he was also groping towards a classification of different types of elements. Further evidence of interest in chemical theory can be seen in his addition to Jabir's analysis of solids into sulphur (inflammable) and mercury (volatile). To this Al-Razi added salt. He considered this third principle a necessary component of solids, as it was neither volatile nor inflammable.

Sadly, alchemy was to cause Al-Razi some tribulation in his old age. In order to ingratiate himself, or perhaps in search of a safe sinecure, Al-Razi wrote a treatise on alchemy which he dedicated and personally presented to the Emir of Khorassan in north-eastern Persia. While reading his splendid gift the emir became so intrigued with alchemy that he summoned Al-Razi and ordered him to conduct a public experiment showing how transmutation worked. Al-Razi hedged, explaining that the apparatus alone would cost a fortune. Regardless, Al-Razi was paid a thousand pieces of gold and told to set up a laboratory for his demonstration. On the appointed day, the emir arrived to witness the transmutation of base metals into gold – complete with Al-Razi's book on the subject, so that he could follow every stage in the process. However, despite hours of dithering amongst the crucibles, refining alembics (distilling apparatus) and furnaces, the old master alchemist inexplicably failed to produce anything which could even be passed off as gold. Whereupon the emir became so enraged that he began to beat Al-Razi about the head with his book. This is said to have been the cause of Al-Razi's blindness during his last years, which he spent in 'poverty and obscurity'. (Curiously, this failure on the great alchemist's part did not deter future generations of adepts, who remained convinced that Al-Razi had finally cracked it.) Al-Razi died in his seventies, some time around 930, and is still deservedly remembered as one of the finest scientists of the Arab world.

Half a century after the death of Al-Razi came the greatest Muslim intellectual of them all – known to us as Avicenna, but in Arabic ibn Sina. Avicenna was perhaps the only man in history to have made major contributions to medicine, philosophy, physics, Arab politics and alchemy. Perhaps not surprisingly, in a figure of such supreme intellect, his most important contribution to alchemy was to doubt its very *raison d'être*: its ability to transmute base metals into gold. Needless to say, this won him few friends in the profession.

Avicenna was born in 980 near Samarkand, the son of a Persian tax collector. Even as an infant he showed exceptional qualities, and by the age of ten, it was said, he could recite the entire Koran from memory. His intellectual abilities soon came to notice. After his education at Isfahan and Tehran, he was employed by various Muslim leaders – sometimes as vizier. This was a dangerous business at the best of times, and even more so during the decline and falling apart of the Arabic empire. Avicenna suffered the usual occupational hazards of political life in the Middle East:

he escaped the death penalty by a whisker on more than one occasion, was kidnapped and held to ransom, and spent several spells in dungeons or in hiding. But, then as now, there were also rewards: Avicenna enjoyed a life of fame, wealth, untold women and, presumably untold, wives. Despite the forbidding of wine by the Koran, Avicenna's consumption of wine was said to have been prodigious.

How he managed to find time for his profound and extensive intellectual activities amongst all this remains a mystery. Perhaps high-living prime ministers didn't even pretend to work so hard in those days.

In his scientific writings Avicenna proposed that a body stays in the same place, or continues moving at the same speed in a straight line, unless it is acted upon by an external force. Here is the first law of motion, set down six hundred years before Newton. Avicenna also pointed out the indissoluble link between time and motion, by means of a compelling poetic image. If every single thing throughout the world was motionless, time would have no meaning. (Not until Einstein was the link between space and time proved mathematically.)

In medicine Avicenna was the most important physician between Galen, the supreme medical mind of the Roman era, and Harvey, who was to discover the circulation of the blood in the seventeenth century. Avicenna's expertise arose directly out of the alchemical lore he absorbed from Al-Razi, and his own alchemical researches. Like Al-Razi, he believed that medicine was a science. In his view, mineral or chemical remedies were far superior to the herbs and old wives' tales that had been prevalent since time immemorial. Avicenna compiled an extensive list of chemicals, their effects when taken as drugs and the diseases which they were capable of curing. This pharmacopoeia was soon accepted as the standard work on the subject.

Avicenna's scientific and philosophical work was eventually

curtailed by political events. He fell from favour as vizier to the Shah of Persia, but managed to escape with his life by going into hiding. He emerged only when the shah became so ill that the court physicians despaired, claiming Avicenna alone could save his life. Avicenna's presence was thus essential, and his safety assured. When the shah was defeated, Avicenna's intellect was considered part of the booty; despite having directed the Persian war effort, he was immediately set to work by the enemy. (An early instance of a tradition which was still flourishing in the Second World War, when the Russians and the Americans both clamoured to get their hands on the best German rocket scientists, regardless of their collusion with the Nazis.)

Meanwhile Avicenna continued as best he could with his philosophical thinking. This, like his chemistry, was grounded on Aristotelian misconceptions. It was further hampered by the conceptual requirements of increasing Islamic orthodoxy. Without this straitjacket Avicenna might well have developed a philosophy of genuine originality. His thirst for philosophical and scientific knowledge was driven by a very modern sense of existential bewilderment, as is shown by his poetry:

> How I wish I could know who I am,
> What it is in the world that I seek.

Despite such claims to ignorance, Avicenna was not one to suffer fools gladly, and his abrasive character won him few friends. He even dismissed the philosophical writings of his medical mentor Al-Razi, suggesting that he should have limited himself 'to testing stools and urine'. Avicenna died in 1037, probably of poisoning.

Within a few years Avicenna's works on philosophy and medicine had achieved widespread currency through the length and breadth of the Arab world. A copy of his pharmacopoeia was found as far afield as the great library at Toledo, when the city was recaptured by the Spanish in 1095. But even before this its secrets had been smuggled to Europe by Constantine of Africa – one of those figures who pop up in history, their name attached to one significant deed and a few suggestive facts, leaving us to imagine the romance of an entire life. Constantine of Africa was born a Muslim, probably in Carthage, and was educated in Baghdad. One day he turned up mysteriously at the medical school in Salerno with a copy of Avicenna's pharmacopoeia. After translating this work into indifferent Latin, he became a Christian monk at Monte Cassino, where he died in 1087. In the following centuries Avicenna's pharmacopoeia was to become the most influential medical text in Europe – the forerunner of modern pharmacy.

3

Genius and Gibberish

As the Arabic empire fragmented and declined, its great contribution to science and mathematics came to an end. This learning now passed to the West, along with many previously unknown Greek texts which had been preserved and used by Arabic thinkers. Only a few works of Aristotle had survived in Europe. As a result of conquests in Spain, and sporadic occupation of the Holy Land by Crusaders, many more of his works began to reappear. These were translated from the Arabic into Latin, which had remained the pan-European language of learning.

These new works arrived in a largely static, hierarchical society, with the Church as its repository of unquestioned values. The earth lay at the centre of the universe, the Pope was God's representative in this world, and the answer to all questions lay with God.

This was not so much an anti-scientific age as an a-scientific one. There was simply no need for science in an age without progress, an age which believed that timeless spiritual values were superior to the vagaries of reality. Innocence or guilt was determined by torture and the ducking-stool, not forensic examination or reasonable argument. The Black Death was combated with prayer, not prophylactics. (When the Black Death swept Europe in the mid fourteenth century it killed over thirty million people – a third of the population of the entire continent.) This horrific event, as much as anything else, accounts for the intellectual stagnation of the period: Europe remained in a traumatized state for almost a century afterwards. Yet, ironically, it was the widespread disruption of the feudal order caused by this plague which opened the way for change.

Scientific thinking was marginalized, with no outlet. Yet, like learning in the so-called Dark Ages, the light of science continued to flicker faintly at the perimeter of this religion-pervaded society. It is often said that the Middle Ages contributed nothing to science. This is not true. Technological advances were few and unspectacular – but significant after their own fashion. Medieval progress is perhaps best exemplified by the invention of the wheelbarrow. More seriously, the horseshoe and the horse collar were developed – as was the mechanical clock. Progress plodded forward at a nag's pace, its motion recorded by the cogs and weights of the town clock, ringing out the hours. These clocks marked gradual astronomical movements in the heavens, sounding the hours for matins and vespers, the opening and shutting of the city gates at sunrise and sunset. Yet despite such timeless tradition, the passing of time was inadvertently moving into an era of smaller and more precise measurement. There was no use for minutes, let alone seconds, yet in order to remain accurate the mechanism of the clock had to measure out such units. It was as if the minutes and seconds were there, accumulating, biding their time for a more hurried, precise age.

In similar fashion, medieval thought was moving inexorably towards an unwonted era of exact examination. The medieval mind accepted the basic premise of science: causation. Everything that happened was the effect of a previous cause – such thinking had been inherited from Aristotle. And it was to be used by Thomas Aquinas, the epitome of medieval theologian-philosophers, as a proof of God's existence. God was the first cause which had set the whole process in motion. This interweaving of science and theology was to prove disastrous. Science moves forward by questioning previous assumptions. Question theology

Alchemist's laboratorium and library

and you are a heretic. Science and God were being set up for an unnecessary conflict, which continues to this day.

In many respects alchemy was made for the medieval mind. Here metaphysics and the world were inextricably entwined: base metals transmuted into gold; the appetites of the flesh transmuted into the strivings of the spirit. But it was dangerous too. Alchemy sought to change the world: to bring base nature to golden perfection, to create order out of chaos. This was God's province – and even to attempt to play God was profanity.

Despite this, one of the first great European figures in alchemy was a priest, who has since been canonized by the Roman Catholic Church and is now the patron saint of scientists. Albertus Magnus was born around 1200 in southern Germany. He studied at Padua and went on to become the finest teacher of his age in Paris. (At the age of twenty Thomas Aquinas walked all the way from southern Italy to become his pupil.)

Such was the state of learning in the early thirteenth century that one could aspire to 'know everything'. Albertus not only took up this challenge, but also sought to extend human knowledge: in philosophy, in what we would call chemistry, and in biology – as well as in the field of alchemy. His vast learning earned him a reputation as a wizard amongst his jealous colleagues. This was undeserved: his approach was scientific (by the standards of the time) and orthodox (he accepted the limiting precepts laid down by Aristotle). In alchemy he was inclined to doubt the possibility of transmuting base metals into gold, though he remained open-minded – perhaps because Aristotle had not pronounced on this matter. He did much skilful pioneering work in his alchemist's den, and his lab notes indicate that he was almost certainly the first to isolate the element arsenic.

Despite its heretical and magical aspects, alchemy appealed to Albertus, and many other genuinely inquiring minds of the period, because it appeared to be a novel and unique way of discovering truth. This was the sole intellectual pursuit which sought to discover truths about the materials of the world. Quite simply, at this period alchemy was the only real science of matter. Hitherto, change had been considered only in its Aristotelian forms – the motion of projectiles, the ageing process, the seasons and so forth. Albertus Magnus was perhaps the first to grasp the idea that chemical change was something entirely different.

Albertus Magnus seems to have lived easily within the constraints of Aristotelianism. His great scientific contemporary Roger Bacon did not. Roger Bacon was born *c.* 1214; he became a Franciscan monk and studied at Oxford and Paris, where he also taught. Like Albertus, his knowledge was exceptional both in its range and its depth. He even attempted to write an encyclopedia which would contain all human knowledge, but he was forced to admit defeat. This was a rare occasion, for Bacon was a man of imperious self-confidence, and was constantly disparaging intellectual failings in others. His was not a monastic temperament: poverty and chastity he could manage intermittently, but obedience was quite beyond him.

Such was Bacon's brilliance that he attracted the attention of Pope Clement IV, who became his patron. When Clement died, Bacon's enemies took their revenge. Eventually the head of the Franciscan order had him imprisoned in Paris for fifteen years, and ordered all his works to be destroyed. Fortunately some were secreted away by like-minded monks, though his *Opus Majus* was not to be published until almost 450 years after his death in 1733.

Bacon's ideas bear a remarkable resemblance to many of those which Leonardo da Vinci sketched in his notebooks – though they pre-date Leonardo by two hundred years, and in many instances go beyond him. Bacon predicted steamships, automobiles, submarines and even flying machines. He suggested that one day people would circumnavigate the globe. One of his letters even contains the first European reference to gunpowder. (As a result, it was thought for many years that he had invented it. Later historians maintained that gunpowder came from China. Recent scholarship suggests that it may well have been invented independently in Europe, in which case Bacon was as likely to have achieved this feat as any.) Such were the exceptional imaginings of an original mind. More mundane, but more significant, was his emphasis on experiment as the only true way forward in science. (Bacon's personality ensured that he was able to spend much time undisturbed in his Oxford laboratory.) He also stressed the application of mathematics as the path to exact truth in scientific experiment. Both these ideas took hold only with Galileo almost four centuries later.

Despite his exceptional scientific vision Bacon remained convinced of the basic premise of alchemy: the possibility of transmuting base metal into gold. Here he succumbed to Aristotelian notions. Aristotle's deepest love had been biology. In nature everything appeared to have its purpose – from the rose's thorn to the cat's whisker. This precept Aristotle and Aristotelian thinkers had adopted for their entire philosophy. The world was purposive: all things strove towards perfection. In humanity the spirit strove to overcome the flesh. Similarly in the world of metals. All base metals strove to become gold. But nature needed some material agent to assist it in this task. Bacon accepted the Arabic notion that an elixir was required as a catalyst to facilitate this process of transmutation. But no record remains of him having successfully experimented with such a substance.

Which brings us to the inevitable question. How on earth did alchemy get away with it for so long? Why wasn't it simply exposed as a hoax? Certainly the obfuscation of the subject by the alchemists themselves helped. Their descriptions of their experiments – in terms of metaphor and mysticism – were impenetrable to the outsider. Likewise, only adepts of the dark art could hope to understand the symbols involved. But more basic forces were at work here. Namely, the will to believe. To say nothing of cupidity and avarice. The blend of mysticism, the initiated, secret knowledge and the prospect of unlimited reward – this heady brew fulfils a primeval need. At the same time alchemy was, after its own fashion, becoming the science of matter. It represented progress in our understanding of the material world. As well as the quest for gold, alchemy was also the quest for knowledge. Had it been exploded as a hoax, and vanished from human knowledge, the entire notion of chemistry would have been lost, perhaps for centuries. Like astrology, alchemy was a wrong turning in human knowledge, a mistake. Yet astrology enabled us to analyse and delineate specific elements of personality, helping us to think about who we are, long before the advent of psychology as a science. Similarly alchemy enabled us to ask – and go on asking – about the material world, to question what precisely it is.

Separating truth from legend is always easy afterwards, when we can apply modern criteria. At the time, the picture was more hazy. What are we to make of Roger Bacon's greatest creation, the one for which he was best remembered in the centuries after his death? According to the story, the visionary who conceived of submarines and aeroplanes constructed a machine of his own. This was a mechanical man with a 'brazen head'. One night while Bacon slept it began to speak, and then shattered. Was Bacon half a millennium ahead of Frankenstein too?

Bacon was to end his days a broken man. In 1291, after fifteen years in his Paris cell, he was eventually released on grounds of health. Now over seventy and seriously ill, he made his way back to Oxford, where he died the following year.

By now alchemy had entered a new creative flowering, its central quest assuming epic proportions. Ironically, this was due in part to a catastrophic loss. In 1204 the French soldiers of the Fourth Crusade overran Constantinople, deposing the emperor. Amidst scenes of mayhem, a drunken prostitute was installed on the emperor's throne beneath the great dome of Hagia Sophia, one of the major churches of Christendom. During the consequent pillaging, countless ancient Greek and Byzantine manuscripts were lost – including a vast heritage of alchemical lore. This was to prove a mixed blessing where alchemy was concerned. It left a huge gap. So instead of slavishly attempting to decipher ancient esoteric texts, the thirteenth-century alchemists were now inspired to undertake their own original endeavours. The result was a significant development in alchemy. The elixir of the Arabs now became transformed into the 'philosophers' stone'. This fabled substance – resonant of mystic quests, the ultimate learning and 'truth' in all its guises – was described by the fourteenth-century Spanish alchemist Arnold of Villanova as follows: 'There exists in Nature a certain pure substance, which when discovered and brought by Art to its perfect state, will convert to perfection all imperfect bodies that it touches.' Others went further. Despite the fact that no one had ever actually seen this elusive stone, it was described as a heavy, glowing powder which exuded a heavenly perfume. When it was red, it could turn base metals into gold; when it was white, it turned them into silver. Like the unicorn, the philosophers' stone had all manner of striking qualities – except existence.

This untoward introduction of the 'philosopher' into alchemy indicates just how far this 'science' had strayed from its original contact with philosophy. One has only to think of Aristotle observing a stone drawn to the earthy bed of a pool, its clinging bubbles then rising through the water to return to air, and meditating on how the four elements are drawn to their natural places. Here were the clarity and precision of the genuine philosopher at work, a far cry from such chimera as the philosophers' stone. But then chemistry was always essentially a murky business. And the quest for the philosophers' stone was to be no different. Amidst the acrid fumes of the alchemist's den and the perfumed clouds of allegory, this quest was liable to lead literally anywhere. Indeed, according to the alchemical text known as *Gloria Mundi*, the philosophers' stone

> is familiar to all, both young and old. It is discovered in the countryside, in the village and in the town, in all things which God has created. Yet it is looked down on by all. Rich and poor alike lay hands on it every day, servants cast it out into the street, and children play with it. No one prizes it, though apart from the human soul it is the most precious thing upon

earth, and can destroy kings and princes. Yet it is regarded as the vilest and meanest of things.

By now the philosophers' stone begins to sound like a riddle, a pretentious literary symbol or even a homily on the perils of lust. Yet, as the nineteenth-century German chemist Justus Liebig put it, 'The finest imagination in the world could not have conceived of a better idea than the philosophers' stone to inspire the minds and faculties of men. Without it, chemistry would not be what it is today. In order to discover that no such thing as the philosophers' stone existed, it was necessary to ransack and analyse every substance known on earth. And in precisely this lay its miraculous influence.'

By the mid fourteenth century copyists in monasteries throughout Europe were reproducing alchemical manuscripts describing the quest for the philosopher's stone. Many of these contained important research into the properties of chemical compounds. But these early inadvertent chemists remain for the most part anonymous. Partly through boredom, partly through ineptitude, the copying monks maintained the tradition of anonymity or false attribution so common in medieval manuscripts; in some cases, too, the original authors themselves were to blame – diffidently ascribing their work to some earlier authority. Some of the most important work of this period was produced by an alchemist who called himself 'Geber', after the eighth-century Arabic alchemist Jabir. This fourteenth-century adept, who was probably a Spanish monk, is now generally known as the 'False Geber'.

Others were less reticent. A few alchemists even published their memoirs, one going so far as to describe his successful quest. In his *Exposition of the Hieroglyphicall Figures* the Parisian scrivener Nicolas Flamel tells how he set out on an alchemical pilgrimage through France and Spain, where he met a certain Master

Canches, a Jewish physician. Before he died, Canches passed on an inkling of the secret of transmutation: 'the first Principles, yet not their first preparation, which is a thing most difficult, above all the things in the world. But in the end I had that also, after long errours of three yeeres, or thereabouts; during which time I did nothing but study and labour.' Finally 'in the yeere of the restoring of mankind, 1382,' he records, 'I made proiection of the Red stone upon the like quantity of Mercurie the five and twentieth day of Aprill . . . about five a clocke in the Evening; which I transmuted truely into almost as much pure Gold, better assuredly than common Golde, more soft, and more plyable'. Sadly, the accompanying explanation of precisely how he did this is couched in the traditional allegorical form, which has remained impenetrable to all eager readers wishing to emulate his scientific Midas touch. Whether or not we believe Flamel, there is no doubting the fact that despite his humble origins he soon became a very rich man, renowned for making generous donations to the churches of Paris. (A contemporary marble plaque recording one of these gifts can still be seen in the Musée de Cluny in Paris.)

But Flamel was not alone in his success. According to tradition, again backed by contemporary records, the Catalan monk Raimondo Lul also succeeded in the 'Great Work', as it came to be known. Lul, better known in English as Raymond Lully, was an ascetic mystic as well as a successful alchemist. With his gold he is said to have paid off the debts of Edward II, the profligate homosexual King of England, much of whose reign was spent combating his homophobic barons. Lully may well have shared Edward's predilection during his own profligate period before he became a monk. Denials of this and of his alchemical bent have become something of a tradition amongst protective scholars – but the fact is that all this, as well as his success as an alchemist, was firmly believed at the time. History is not what actually took place, but what we believe took place. Despite widespread charlatanry there is no doubt that many alchemists of this period were honest in their endeavours. They believed utterly in what they were doing. And they had good grounds for doing so. Out of Aristotle's original philosophy, medieval philosophers had developed their own Aristotelianism. According to this, Nature's striving for perfection also took place amongst the minerals beneath the earth's surface. Stones became rocks, rocks became metals. A process of evolution was constantly taking place. Over the years base metals in their striving slowly ascended the ladder of perfection – becoming tin, then silver and eventually gold.

Such thinking is not as far-fetched as it might seem. When smelted, earthy ore produces pure metal. And this was a simple process compared to other transformations observed by medieval scientists. How much more difficult was it to explain the apparent transformation of rotting meat into worms, or caterpillars into butterflies, seeds into towering trees? Here was transmutation indeed. By comparison, the transmutation of one metal into another by evolution was a simple, easily credible process.

Yet simultaneously these same alchemists were also laying the theoretical foundations of chemistry as we know it. In their ravening search for the philosophers' stone the fourteenth-century alchemists became the first to understand the nature of acids. The only acid known to the ancients had been the weak acetic acid of vinegar. In the eighth century Jabir had prepared a weak solution of nitric acid, and other Arab alchemists discovered that distilling vinegar made a stronger acetic acid. But even strong acetic acid was hardly corrosive. It didn't appear to be possessed of much reactive power. Then came the breakthrough. Just after 1300 the False Geber discovered vitriol, better known to us as sulphuric acid. Here was a liquid which appeared to dissolve, corrode or react with almost everything! This has been called

Alchemy

the most significant chemical advance since the discovery of how to produce iron from its ore, around three thousand years previously. In time sulphuric acid would similarly transform the world. (Until the mid twentieth century the index of a country's development was measured by the volume of sulphuric acid its industry used each year.)

Besides sulphuric acid, False Geber also described how to make strong nitric acid – which was called aqua fortis, literally 'strong water', because of its ability to dissolve almost anything except gold. Previously most alchemists' use of acids had been confined to those weak acids which occurred naturally – such as acetic acid from vinegar, or lactic acid from soured milk. The discovery of how to obtain strong acids from minerals opened up an entire new vista of experiment. Substances could be dissolved in mineral acids, metals were corroded by them to form salts, solutions formed precipitates when they were added. The alchemists had stumbled upon the means which would enable them to perform a vast range of basic chemical reactions – the forming and dissolution of compounds, the transforming of one compound into another. Likewise, they had now discovered a method of isolating elements previously found only as compounds. But all this we can see only with hindsight. The alchemists had no real theoretical framework on which to collate the results of their experiments. They were blundering in the dark. All they had discovered was the practical means. They knew how to do it – but they didn't really know what they were doing.

Unfortunately, the alchemists themselves were convinced otherwise. They knew exactly what they were doing. Their experiments were performed with but one objective: the transmutation of base metals into gold. Chemical discoveries were harnessed to this end, or ignored. Little systematic exploration of these new possibilities was undertaken – what was the point?

Despite this blinkered approach, systematic progress of a sort was achieved, if somewhat inadvertently. The key to this lay in the notion of elixirs. Medieval Europe inherited a muddled idea of elixirs from the Arabs. The elixir was the catalyst which induced the transmutation process; but the notion that this substance must be magical in itself persisted. The elixir was also the elixir of life, which bestowed immortality. By association, it soon came to be regarded as having medicinal qualities. This conceptual ambiguity is exemplified in the writings of John of Rupescissa, who practised in the mid fourteenth century. Little is known of his life, apart from the fact that he was imprisoned for casting aspersions on the moral excellence of Pope Innocent VI, ironically one of the better-behaved popes of the period. John of Rupescissa's writings, on the other hand, are somewhat less forthcoming. In some places, the elixir he names as a catalyst for transmutation is almost identical to an elixir he prescribes for some medical complaint. Purging lead of its base elements and purging the bowels seem to have been viewed as much the same process – both requiring identically drastic treatment. The effects of swallowing a catalyst designed to operate in molten metal at around 400°C can hardly be imagined.

Regardless of any protests from the patients, this was to be the start of a trend. In order to generate some much-needed income, alchemists soon began prescribing different elixirs for different medical complaints. Soon specific elixirs became recognized as appropriate treatment for particular complaints. The idea of using chemical substances to cure specific complaints had been central to Jabir's pharmacopoeia – copies of which, in Constantine of Africa's translation, had now spread widely through Europe. The elixirs produced by the European alchemists reinforced this idea. Pharmacy was being born in Europe.

Such practice was aided by the discovery of the most important elixir of all time. This was yet another strong water, or *aqua*; it became known as the water of life, or aqua vitae. It was produced by the careful distillation of wine. The alchemist who first produced almost pure alcohol was Arnold of Villanova, born in Spain in the fourteenth century. Arnold's thinking was a curious blend of mysticism and scientific insight. He observed that when wood is burnt in an unventilated room there is an accumulation of poisonous gas – making him the discoverer of carbon monoxide. He also believed that the philosophers' stone existed in all substances, from which it could be drawn out. This mystical idea echoes his preparation of pure alcohol by the distillation of wine. As a result of alchemical symbolism, alcohol came to be regarded as the essence of sunbeams (aetherial gold) which had pierced the grape and been retained in its juices. (It looks as if wine writing was also invented in this period.)

Arnold of Villanova was a man of wide medical learning who knew Arabic. He read Avicenna's pharmacopoeia in the original, rather than Constantine of Africa's somewhat haphazard translation, and his prescriptions were correspondingly more precise than those of his professional rivals. Throughout southern Europe, Arnold of Villanova was recognized as the finest physician of his day, summoned to the bedsides of royalty and popes (of whom there were nearly twenty during his lifetime). Treating ill, irascible monarchs and over-indulgent pontiffs was a perilous business. Failure could be fatal for both the patient and his doctor. But success brought similarly exciting rewards. Arnold received a magnificent castle from King Pedro III of Aragon, a well-paid professorship at Montpellier University from a pope, and country estates in Italy, Spain and the south of France – all ideal spots for him to continue his researches into the distilling of wine.

Like aqua fortis (nitric acid), aqua vitae (alcohol) was found to be a solvent, though of a different, more subtle tenor. It also had preservative qualities. In medicine it could be used as a disinfectant, to clean wounds. Others, less philanthropically minded, put this miraculous new liquid to their own peculiar experimental use. This habit quickly spread throughout Europe, where 'aqua vitae' was translated into several different languages. In French it became *eau de vie*, in Scandinavian languages *akvavit*, in Gaelic *usquebaugh* (whisky). Such experiments must have proved inconclusive, for they continue to this day.

Almost as a by-product, alchemists were coming up with important scientific ideas. Meanwhile alchemy itself continued to come up with little more than fool's gold. Not surprisingly, the dark art now entered a third period of decline – the previous ones being at the end of the Roman Empire and the break-up of the Arab Empire. Once again, alchemy was officially banned: this time by Pope John XXII in 1317. (Though not until after he had spent several years attempting the great work himself, having received instruction from no less a figure than Arnold of Villanova. Opinion at the time was divided over John XXII's motives for banning alchemy. Some said it was due to disillusionment, others claimed it was pique at his lack of ability. Still others claimed that he wished to monopolize the field, and that he continued to practise alchemy in secret in the dungeons of the papal palace at Avignon. They maintained their suspicions were vindicated when Pope John XXII died leaving an unaccountable fortune in gold, worth eighteen million francs. Metaphysicians and accountants continued to dispute the source of this vast sum for many years.)

But the papal ban merely drove alchemy underground. The dark art had struck a chord in human nature – gold, secrecy, esoteric learning: the combination remained an unfailing attraction to seekers after knowledge. Even the greatest medieval mind of them all, the theologian-philosopher Thomas Aquinas, was interested. Having received instruction from his teacher Albertus Magnus, he is said to have written a work called *Thesaurus Alchemiae*, as well as several similar tracts. The authorship of these works remains disputed. Yet according to the historian of alchemy A. E. Waite: 'Some of the terms still employed by modern chemists occur for the first time in these suppositious writings of Thomas Aquinas – e.g. the word amalgam, which is used to denote a compound of mercury and another metal.' It is difficult to imagine how the woolly ramblings of hermetic philosophy could have coexisted with the theological rigour of scholastic philosophy in the mind of Aquinas. Yet ironically these very same ramblings were to have a fundamental effect on the European mind and what might be termed its Philosophy of Life.

This can best be illustrated by the ancient alchemical text known as *The Emerald Tablet*, written by the mysterious Hermes Trismegistos. This work, which was probably written in Alexandria around the first century AD and had later become central to Arabic alchemy, probably began to circulate in Europe some time in the fifteenth century. It is believed to have been one of the many ancient books brought to the West by Greek scholars fleeing Constantinople some time before the capital of the Byzantine Empire fell to the Ottoman Turks in 1453. This exodus contributed significantly to the rebirth of classical learning which flowered into the Renaissance, but it also contributed much metaphysical dross.

The Emerald Tablet contains many prime examples of the usual mumbo-jumbo. 'Whatever is below is like that which is above, and that which is above is like that which is below, to accomplish the miracle of one thing.' But amongst all this, it also contains a very definite credo. According to *The Emerald Tablet*, God created man much more closely in his image than had previously been realized. Besides being a rational soul, man was also a creator. But in order to exercise this divine power, he first had to discover the secrets of nature. This could be done only by torturing nature – by subjecting it to fire, dissolving it with strong waters and subjecting it to other alchemical processes such as distillation.

Success in these manipulations turned man into a god possessed of eternal life. It gave him power over matter, untold riches and the ability to turn the world into a paradise. It is not difficult to recognize in this the vision which was to evolve into the central belief of science. Through technological innovation, experiment and scientific thinking man could rule nature, transforming it, imposing his will upon the world. According to the historian of science L. Pearce Williams, 'This is essentially the modern view of science, and it should be emphasized that it occurs only in Western civilization. It is probably this attitude that permitted the West to surpass the East, after centuries of inferiority, in the exploitation of the physical world.' In the midst of the most recalcitrant alchemy lay the seeds of the scientific fantasy which continues to propel the world today (whether we like it or not).

Scientific thought was coming full circle. Having started with Greek philosophy, it was now returning to this original clarity. On the way, it had picked up many unscientific notions. Some of these would not be so easily shed, and some even proved positively useful. *The Emerald Tablet* is permeated with mystical ideas and correspondences. The original seven elements are related to the seven planets, which in turn control the seven days of the week. Gold is related to the sun, whose day is Sunday. Silver is related to the moon, whose day is Monday. The effects of such hermetic notions still lurk in our calendar. Other such notions played their part in even the greatest scientific ideas. *The Emerald Tablet* incorporated much late Platonic mysticism, including the notion that the sun was the source of our enlightenment, spiritual as well as physical. This derives from Plato's famous parable of the cave. According to this, we exist like bound prisoners in a dark cave. There is a fire at our backs, which casts shadows on to the cave wall in front of us. This world of appearances we mistake for reality. Only if we learn to turn away from this world of illusion can we begin to see the real light of the sun outside the cave. The sun is the true reality and is central to our world.

When Copernicus translated this idea into scientific fact, announcing that the earth and the planets orbit the sun, his inspiration was not entirely scientific. In his *De Revolutionibus Orbium Coelestium* ('On the Revolutions of the Celestial Spheres'), he specifically mentions Hermes Trismegistos, using him as an authority to back up his revolutionary idea. Throughout this work Copernicus' language remains highly Platonic in tone. Indeed, it was the influence of Plato that was responsible for his major error. According to Plato, the heavenly bodies (i.e. the planets) could only obey 'real' geometry, which was limited to such ideal figures as could be drawn with a compass and ruler. It was this which led Copernicus to assume that the orbits of the planets must be precisely circular; only later was it realized that they must in fact be elliptical.

Science was entering a new era of discovery which would change for ever the way we view the world. Yet at the same time it remained attached to old ways of thinking, which could affect its entire vision. These contradictory forces were to be embodied in one of Copernicus' most notorious contemporaries, who in many ways epitomized this era of science.

4

Paracelsus

Theophrastus Bombast von Hohenheim, better known to history as Paracelsus, was born in the Swiss village of Einsiedeln late in 1493. Just a year earlier Columbus had reached America and Lorenzo the Magnificent had died in Renaissance Florence. Gunpowder was transforming warfare, rendering even the most impregnable feudal castles vulnerable; double-entry book-keeping had transformed banking, enabling it to finance and audit large-scale commercial enterprises; and Gutenberg's invention, the printing press, was now in use throughout the West, from England to southern Italy. Europe stood on the brink of a new historical era.

Not for nothing was Paracelsus' middle name Bombast – for many years the present meaning of the word was thought to be derived from his name, and there's no denying that his behaviour gave added impetus to this meaning. But in the beginning things were very different. Paracelsus was a sickly child, and suffered from rickets. His father was illegitimate and his mother had been a bondswoman (a virtual slave, owned by her employer). This background of physical and social deprivation played a formative role in Paracelsus' complex and antagonistic personality. He was also said to have been emasculated in infancy. How or why this happened is uncertain – it was possibly due to illness. Throughout his life he would remain beardless, and his features had a certain effeminacy. He showed no interest in sex, and masked his deficiencies by adopting a boisterous, laddish penchant for carousing.

Paracelus, aged forty-seven

Paracelsus' mother died when he was a child, and he moved with his father to Villach in Austria. The wonders of this 250 mile journey on foot through the Alps at a formative age must also have had their effect. Paracelsus was to remain footloose throughout his life. At Villach Paracelsus' father taught practical and theoretical alchemy at the local mining college. However, his role was not that of some tolerated wizard, held in the same regard as, say, a professor of paranormal psychology at a modern university. The subject he taught would now be called practical and theoretical metallurgy. Yet there's no doubt that Paracelsus senior tried his hand at the odd transmutation in his spare time, with young Paracelsus as his assistant in the smoky alchemist's den. Such practice would have been neither hypocritical nor illogical. The theory of metallurgy in those days still encompassed the belief that metals were 'refined' in the ground, gradually evolving from the baser forms into silver, and finally into gold. Alchemical transmutation was merely the scientific attempt to speed up this process.

Paracelsus' early experience with his father made him highly expert in both the properties and the handling of minerals. This expertise was extended when he began working, perhaps as an apprentice overseer, in the local mines and workshops owned by Sigismund Fugger, who was also a keen alchemist. Fugger was a member of the great German mercantile family which played a central role in European commerce during the fifteenth and sixteenth centuries. The Fugger family had mining interests from Hungary to Spain, and a network of banking agents extending from Iceland to the Levant. The family even accumulated sufficient wealth to finance such operations as the wholesale bribery necessary to ensure that Charles V rather than Francis I of France became Holy Roman Emperor. (The Fuggers were pre-eminent in northern and eastern Europe; meanwhile a comparable role in southern and western Europe was played by the more cultured and longer-lasting Medici banking dynasty, who in certain aspects superseded the Fuggers.) Working in the Fugger mines Paracelsus learned a crucial lesson he would never forget. Here alchemical theory always remained second to practice. Success was judged by production.

Paracelsus' employment by Fugger seems to have ended around 1507, when Paracelsus was just fourteen, and he then set off on foot to seek learning at the universities of Europe. This was not unusual in medieval times, but Paracelsus' exceptional youth indicates his wilful temperament as much as his prodigious intellect. For the next few years he lived the life of a wandering scholar. At Württemberg he attended lectures by Trithemius, the alchemist-astrologer who invented one of the first widely accepted shorthand notations. In Paris he studied under Ambroise Paré, the army surgeon who first learned to tie human arteries, now recognized as the father of modern surgery. Paracelsus' arrogance and talent for picking quarrels ensured that he never lasted in one place for long. In later years he would claim to have taken his degree at an Italian university – though precisely where remains uncertain. And he invariably listed amongst his qualifications the doctorate in medicine he gained at Ferrara in 1517. (Though historians have been unable to check on Paracelsus' claim here, as the Ferrara University records for this year are missing, a fact of which Paracelsus was probably well aware.)

It was during his early twenties in northern Italy that Paracelsus began to proclaim for the first time his unorthodox academic ideas. The universities of Europe remained firmly rooted in the medieval past, despite the birth of Renaissance humanism – with its new emphasis on human (rather than spiritual) values, and on the worth and dignity of human beings. Lectures at the universities were still delivered in Latin, and all disputes were settled by appeal to the texts of the classical authorities, rather than anything so mundane as reality or human experience. And who were these authorities? Aristotle's influence had by now begun to exert a stranglehold on learning, rendering any progress all but impossible. And in medicine Galen and Avicenna held similar sway. Galen, the great physician of the Roman era, had cured emperors in his time, but he had based his knowledge of the finer points of human anatomy upon his dissections of dogs and pigs; Avicenna's first steps towards prescriptive medicine were regarded as the last word on the subject.

Paracelsus would have none of this. He soon found himself rejecting academic teaching altogether. There was only one way to learn medicine. 'A doctor must seek out old wives, gipsies, sorcerers, wandering tribes, old robbers and such outlaws and take lessons from them. A doctor must be a traveller . . . Knowledge is experience.' Practising what he preached, Paracelsus once again took to the road. Yet despite his proclaimed aims, this was the action of no humble seeker after knowledge. Previously he had been known by the name he had been given at birth: Theophrastus von Hohenheim. But now the man whose middle name was Bombast adopted the name Paracelsus – which means 'greater than Celsus'. This was a reference to the first-century AD Roman physician whose recently rediscovered works had become all the rage in academic circles. Paracelsus had recognized at once that Celsus' work was a largely unoriginal regurgitation of earlier Greek sources, especially Hippocrates, the 'father of medicine', who had died in the fourth century BC. But the world had advanced in the previous two millennia, and this glorification of such pedigree classical ideas only goaded the déclassé young bastard's son. He was better than any Celsus, and had every right to call himself so. He'd show them.

For the next seven years Paracelsus wandered the highways and byways of Europe. Sometimes he supported himself as an army surgeon, sometimes he practised as an itinerant doctor. Occasionally he would be called in to minister to some ailing nobleman, and was richly rewarded for his troubles; at other times he was reduced to peddling home-made remedies in the market place. In this way he travelled to the Netherlands, on to Scotland, then to Russia, and as far as Constantinople. Never afraid to embellish the truth, he claimed to have travelled even further afield, to Egypt, the Holy Land and Persia. But later, in an uncharacteristic moment of modesty, he admitted: 'I visited neither Asia nor Africa although it has been so reported.' And everywhere he went he acquired local knowledge – especially of chemical compounds and their effects, old wives' cures, alchemical practices. And everywhere he went they remembered him. According to one contemporary: 'He lived like a pig and looked like a sheep drover. He found his greatest pleasure amongst the company of the most dissolute rabble and spent most of his time drunk.' But others who met him were so impressed they called him 'the German Hermes', 'noble and beloved prince of learning', 'the king of all knowledge'. Paracelsus would always inspire such contradictory responses, and his teachings curiously mirror this. The uncouth, loud-mouthed figure was undoubtedly a charlatan, a market-place quack; but his legacy of scientific ideas, though sometimes inarticulate and often unoriginal, pointed the way to the future. These were the first steps out of the medieval quagmire on to the firm ground of scientific method. Perhaps the man was necessary for the work.

Take, for instance, his visit to Constantinople in 1522. The Ottoman capital during the reign of Suleiman the Magnificent was no place for an infidel. Even ambassadors sent from Europe were customarily flung into the horrific Yedikule dungeons as spies. (Which is of course what diplomats have always been, but Suleiman the Magnificent was not interested in the niceties of diplomacy.) Yet it was here in Constantinople that Paracelsus rediscovered some of the lost secrets of Byzantine alchemy, bringing them to Europe for the first time. He also picked up some equally sensational genuine scientific knowledge. The peasant women practised a primitive form of medicine which seemed to prevent such diseases as smallpox. They would slice open a vein and insert a needle which had been infected with smallpox. Later, Paracelsus was the first European physician to state that, when introduced into the body in small doses, 'what makes a man ill also cures him'. Yet, as ever, the Paracelsian legacy here was double-edged. Some have seen this as an understanding of the principle of inoculation some two centuries before it was understood elsewhere; others have seen it as anticipating the more controversial practices of homoeopathic medicine.

In 1521 Paracelsus was present at the Diet of Worms, where Martin Luther was called upon to defend his attacks on the Catholic Church. Paracelsus recognized a kindred spirit in the headstrong miner's son who defied the Pope and ridiculed the selling of indulgences as 'tickets to heaven'. When Luther refused to recant, insisting, 'Here I stand; I can do no other,' he sparked the Protestantism that would split Europe. Paracelsus was moved to see his own mission in a similar light. Yet he remained a Catholic, if an unorthodox one, and always steered clear of religious controversy. His battleground would be science.

But apart from protest, what exactly was it that Paracelsus wished to convey to the world? The discovery of new scientific methods (or old ones, practised amongst the people, which had long been ignored by respectable knowledge)? A preference for experience rather than authority? An insistence upon what worked? These were certainly different, in their own way – but they hardly constituted any coherent doctrine.

Paracelsus soon put this right by proclaiming as his own the idea of iatrochemistry. This wasn't actually his idea, but he did so much to advance it that he virtually made it his own. Iatrochemistry stems from the Greek *iatros*, meaning physician, and its aim was to establish chemistry at the heart of medical practice.

In effect, chemistry was of course still alchemy. But Paracelsus roundly declared that alchemy was wasting its time in trying to produce gold. The techniques of alchemy should be put at the service of medicine – to produce chemical cures for sickness and diseases, specific medicines being prepared for the treatment of specific diseases. Medicine would then become a science, rather than the faintly dubious art which it then appeared to be. Medical knowledge could be written down in books and these could be consulted by all physicians, who could then prepare any medicine which they required. Each medicine would have a clear, universally accepted name, and would be accompanied by a simple, clear description of how it was prepared. There was no room for the ambiguity and metaphor which bedevilled alchemical textbooks, due to redundant theories. From now on the emphasis would be on practice and the scientific application of prepared cures. There was no need either for theory or for old wives' tales and herbal cures.

There's no denying that Paracelsus was contradicting himself, on two scores. First, he still believed that alchemy would one day produce gold, and throughout his life undertook repeated experiments to this end. Also, he remained a firm believer in old wives' tales about herbal and plant cures practised amongst the people. The distinction here seems to be that these were acceptable so long as he brought them to notice, but not when they were used by some rival physician.

As ever, Paracelsus' insights were a blend of the old and the new, the daring and the daft. Minerals should be investigated in a comprehensive manner to discover their properties; only then would it be possible to assign them as appropriate cures for specific diseases. He insisted that compounds should be prepared in precise fashion from pure chemicals, not mixed together any old how, as tended to be the alchemists' way. This insistence would later lead to one of the most important insights in the history of chemistry. Indeed, it is so important that it seems obvious to us – but it remained unrealized in Paracelsus' day. It was not understood then that the properties of chemical compounds were affected by the elements from which they were made. Such a fact brings forcibly home to us the extent to which the embryo chemistry of Paracelsus' period remained in the dark. Alchemy had brought chemistry only to the point where it was beginning to scratch the surface. The stuff of reality – matter itself – remained for the most part an unknown quantity.

Once again Paracelsus practised what he preached. His study of chemical compounds was thorough, and to this day the modern pharmacy remains well stocked with compounds he investigated and prescribed: zinc and copper salts, compounds of lead and magnesium, arsenic preparations for skin complaints, and so forth. Yet at the same time Paracelsus insisted on deriding advances made by his rivals. Dissection was dismissed as 'dead anatomy'; the effect of the inner workings of the body on diseases was similarly ridiculed (except in the cases which he alone demonstrated).

Despite his dabblings in alchemy, Paracelsus was undeniably an emergent chemist. He believed that all life was in reality a series of chemical processes. The body was nothing more than a chemical laboratory. When it became ill, this was due to a chemical imbalance or malfunction. This could be rectified by introducing counterbalancing chemicals or initiating an appropriate chemical response. So far so good. But once again a far-seeing chemist was hamstrung by backward theory. Paracelsus' view of the elements was a variant on traditional ideas. He accepted Aristotle's earth, air, fire and water. He also accepted the Arabic development of the three principles: sulphur (giving flammability or combustion), mercury (giving volatility and its opposite) and salt (giving solidity). Paracelsus saw these principles as fundamental, and justified them by recourse to the traditional description of how wood burns in a fire. Mercury included the cohesive principle, so when it left in smoke the wood fell apart. Smoke represented the volatility (the mercury principle), the heat-giving flames represented flammability (sulphur) and the remnant ash represented solidity (salt). These three principles were not matter (that consisted of the four elements), they were how matter worked.

In 1524 Paracelsus finally arrived back home in Villach, complete with a large sword which he claimed to have acquired while serving in the Venetian army. (From now on, he would take this sword wherever he went; it became his talisman, his trademark. Many portraits of Paracelsus feature this telling Freudian symbol, which he is even said to have worn in bed.) Paracelsus was welcomed home by his father, who was now a respected senior citizen of the local community. By this stage, largely through constant self-advertisement, Paracelsus had begun to acquire a certain renown. But as his fame grew, so did his head. The hardened traveller-wiseman soon felt that he could do what he liked and say what he liked. At Salzburg in 1525 he was lucky to escape with his life, after publicly voicing his support for the Peasants' War. The tales of his travels are littered with such episodes – some are doubtless exaggerations, but many have a ring of truth, suggesting they may originally have been based on genuine incidents. Heroic drinking bouts and reckless protests against authority are the main recurring themes.

In 1527 Paracelsus turned up in Basel. Here an influential local figure called Johan Frobenius summoned him as a last resort to treat his disabled right leg. The local physicians were all for amputation, but Paracelsus managed to dissuade them from this drastic course just in time. He then effected a spectacular cure. Precisely how he did this remains uncertain, but there's no doubting the fact that he did.

Paracelsus had made an important friend. Frobenius was a wealthy publisher with humanist views whose business had brought him into contact with many of the leading intellectual lights of Europe. At the time when Paracelsus was treating Frobenius, the Dutch Renaissance scholar Erasmus happened to be staying in his house.

Erasmus was a man of vast learning, who prided himself on the independence of his outlook. This was little influenced by the teachings of the Catholic Church; likewise it did not coincide with those often narrow-minded reformers who allied themselves with Luther. Erasmus' non-partisan attitude helped clear the way for the ground-breaking advances in philosophy, mathematics and science which were to take place during the next hundred years. Aware of his own shortcomings, it was he who coined the phrase: 'In the country of the blind, the one-eyed man is king.' The relevance of this saying to Paracelsus' situation is all too evident – though it would not have been acknowledged by the man himself.

Erasmus and Paracelsus took an immediate liking to one another. It is difficult to see what the ailing elderly scholar and the weather-beaten travelling alchemist with a penchant for roistering would have had in common. Erasmus may well have sat 'with wrinkled brow and glazed eye' through the enthusiastic iatrochemical ramblings of his raggedly dressed friend, but he evidently didn't doubt their veracity. He asked Paracelsus to treat him for gout and a painful kidney complaint, for which Paracelsus duly came up with a cure. Erasmus was so impressed by Paracelsus, both as a physician and as a scholar, that he wrote to him, 'I cannot offer thee a fee equal to thine art and learning.' No mean compliment from the finest mind of the age. Erasmus even offered to help Paracelsus find a job worthy of his talents.

On the recommendation of Erasmus and Frobenius, Paracelsus was given the post of town medical officer and lecturer in medicine at the University of Basel. He was now just thirty-three. Bald, his broad, coarse features softened by his lack of beard and faintly hermaphroditic air, he was beginning to show his age: the hard travelling and hard living had left their mark. Yet now at last he stood poised for great things – so long as he behaved himself.

But that was the snag. Paracelsus was congenitally incapable of such a thing. At the beginning of term he pinned his programme of forthcoming lectures to the noticeboard outside the university gate. He announced that contrary to tradition his lectures would be open to all, and would be delivered in German so they could be understood by all. Even local alchemists and lowly barber-surgeons were invited to attend. There was inevitable outrage from the authorities, who recognized at once the inspiration behind this gesture. Ten years earlier, Luther had nailed his 'Ninety-five Theses' against indulgences and the false claims of the Catholic Church to the door of the church in Wittenberg. Luther too no longer preached in Latin, preferring the plain honest German of the people. 'Why do you call me a medical Luther?' Paracelsus asked the authorities disingenuously. 'You wish us both in the fire.' But he knew that his views challenged only medical orthodoxy – there was no risk of his being burnt at the stake. What he sought was the heroic role. He wished to be notorious throughout Europe, and his behaviour at Basel would ensure this.

Despite their sensational billing, Paracelsus' lectures themselves proved no anticlimax. Forswearing academic robes for his opening lecture, he turned up in his alchemist's leather apron. The hall was packed – with students, townsfolk, academics and local physicians come along to see what tomfoolery this maverick would produce. They were not disappointed. Paracelsus opened by announcing that he would now reveal the greatest secret in medical science. Whereupon he dramatically uncovered a pan of excrement. The physicians began leaving the hall in disgust, while Paracelsus shouted after them, 'If you will not hear the mysteries of putrefactive fermentation, you are unworthy of the name of physicians.' Paracelsus believed that fermentation was the most important chemical process which took place in the laboratory of the human body.

Paracelsus went on to dismiss out of hand the prevailing academic theory of bodily health. (This time it was the academics' turn to leave en masse.) The orthodox medicine of the day operated according to the theory of the 'four humours', which derived from Hippocrates. These were blood, phlegm, choler (yellow bile) and melancholy (black bile). The balance of these humours in the body governed its physical and mental qualities. They were responsible for both a person's temperament and his complexion. A choleric person was quick to anger and liable to have a yellow face (especially when he was enraged and the blood drained from his features). When blood predominated, a person was liable to have a sanguine temperament (from *sanguis*: Latin, blood) and a red face. The fact that we still refer to people as sanguine, phlegmatic or filled with bile is a remnant of this theory.

Each humour had its seat in one of the major organs: the heart (blood), the brain (phlegm), the liver (choler) and the spleen (melancholy). The four humours were clearly derived from the four elements – blood (fire), black bile or melancholy (earth), and so on. When a person became ill, it was because he or she was suffering from an imbalance of the humours. For instance, a patient with a fever had an excess of heat, or fire. This element corresponded to the humour blood, so to remedy the fever the patient would be bled, thus reducing the heat in his body. (Such was the compelling power of this ingenious theory that blood-sucking leeches were to remain a part of the physician's therapeutic paraphernalia for well over two thousand years – reaching a high point in the nineteenth century, long after the theory of the four humours had been discredited. No visiting Victorian doctor worth his salt arrived without leeches in his portmanteau.)

The four humours were also connected in a holistic way with the four seasons, the four ages of man and the four points of the compass. They were even controlled by the four major planets: the moon, Mars, Jupiter and Saturn (to this day a person prone to melancholy is sometimes described as having a saturnine temperament). In this way medicine, astronomy, astrology, psychology and alchemy were all bound together in a richly resonant symbolic world.

But Paracelsus would have none of it. Humanity had to be released from this cage of metaphysical symbolism and set free into the open air of actuality. Heaven and earth were not matched, and humanity did not stand at the centre of things, reflecting both. (At the same time as Paracelsus was delivering his lectures in Switzerland, halfway across Europe in Poland Copernicus was also displacing humanity from the centre of things with his heliocentric system of the planets. Though he would be afraid to publish his theory until he lay on his deathbed sixteen years later.)

The truth, according to Paracelcus, lay in iatrochemistry. Disease should be treated with appropriate medicines, which could be prepared from mineral sources. The hard facts of iatrochemistry would replace the vapid 'humours', which remained tied to the four elements of Aristotle. With hindsight, the significance of this shift for chemistry appears crucial. The focus would now be on investigating the actual and multifarious properties of chemical compounds, rather than on ascribing chemical qualities to the balance of four elements in a substance. Instead of four notes, chemistry would now have an entire keyboard to play upon.

So Paracelsus becomes the father of modern chemistry? Alas, no. He was far too ebullient and complex to limit himself to the purely scientific. In a classic example of intellectual schizophrenia, Paracelsus also invented a theory which now appears as the epitome of quaint medievalism. And what's more, he believed this theory, and used it for medical purposes, even though it completely contradicted his ground-breaking scientific theories.

Paracelsus 'doctrine of signatures' claimed the superior wisdom of nature. It was the physicians' duty to seek to understand nature's language, which indicated in simple form how to bring about particular cures. Plants had signatures, which the physician must learn to read. For instance, an orchid resembled a testicle – which meant that it was a cure for venereal diseases; lilac leaves were heart-shaped, and thus good for heart diseases; 'yellow-blooded' celandine was the remedy for jaundice; and so forth.

If we are to believe Paracelsus, he learned this wisdom 'from the peasants', though even his contemporaries were sceptical concerning this source, believing that the peasants had more sense. The doctrine of signatures has all the hallmarks of one of Paracelsus' higher flights of perversity. Not only would he attack the medievalists – he would show that he could go one better than them.

And attack them he certainly did. Just three weeks after lifting the lid from the pan of his first lecture, Paracelsus led a crowd of cheering students to the St John's fair, which was being held in the square in front of the university. Here he proceeded to burn publicly the works of Galen and Avicenna – sprinkling sulphur and nitre on the flames. Some commentators have claimed this was symbolic: Paracelsus intended to demonstrate his faith in the power of chemicals over the redundant works of the past. Anyone who has ever thrown a firework containing these two substances into a bonfire will understand that Paracelsus' action was nothing of the sort. What he was after was spectacular effect. Hellfire and damnation. 'If your physicians only knew that their prince Galen . . . was sticking in hell, from whence he has sent letters to me, they would make the sign of the cross upon themselves with a fox's tail. In the same way your Avicenna sits in the vestibule of the infernal portal.' Here for the first time we hear the authentic rant of the charlatan-genius. Paracelsus was seldom reticent in the face of his enemies: 'Woe for your necks on the day of judgement! I know that the monarchy will be mine. Mine too will be the honour and the glory. Not that I praise myself: Nature praises me.' And behind the bonfire and the ranting there was a less explicit message. The suitably outraged authorities saw only too clearly what Paracelsus was up to. Not six years previously, at the gates of Wittenberg, Luther had publicly burnt the papal bull from Leo X threatening him with excommunication. So much for Paracelsus disclaiming the title 'a medical Luther'.

Paracelsus revelled in his newly aquired position, and felt no compunction about saying precisely what he felt to be the truth: 'All the universities and all the ancient writers put together have less talent than my arse.' The students worshipped him, his lectures became a public event, and his drinking in the taverns became a public outrage. Miraculously, he also continued to produce original work. His secretary Oporinus remembered:

> He spent his time on drinking and gluttony, day and night. He could not be found sober an hour or two together . . . Nevertheless, when he was most drunk and came home to dictate to me, he was so consistent and logical that a sober man could not have improved upon his manuscripts.

Oporinus seems to have had a difficult time with his master:

> All night, as long as I stayed with him, he never undressed, which I attributed to his drunkenness. Often he would come home tipsy, after midnight, throw himself on his bed in his clothes wearing his sword, which he said he had obtained from a hangman. He had hardly time to fall asleep when he rose, drew his sword like a madman, threw it on the ground or

> against the wall, so that sometimes I was afraid he would kill me.

At work it was even worse:

> His kitchen blazed with constant fire; his alcali, oleum sublimati, rex praecipitae, arsenic oil, crocus martis, or his miraculous opoldeltoch or God knows what concoction. Once he nearly killed me. He told me to look at the spirit in his alembic and pushed my nose close to it so that the smoke came into my mouth and nose. I fainted from the virulent vapour . . . He pretended that he could prophesy great things and knew great secrets and mysteries. So I never dared to peep into his affairs, for I was scared.

It's worth quoting the testament of the long-suffering Oporinus at some length. All aspects of his master's character shine through:

> He was a spendthrift, so that sometimes he had not a penny left, yet the next day he would show me a full purse. I often wondered where he got it. Every month he had a new coat made for him, and gave away his old one to the first comer; but usually it was so dirty I never wanted one.

Paracelsus may on occasion have identified with Luther, but his reference to Nature praising him, coupled with his assertion that on the day of judgement 'the monarchy' would be his, as well as 'the honour and the glory', hint at even grander delusions. His attitude to religion is thus hardly surprising:

> I never heard him pray or inquire after the evangelical doctrine which then was practised in our town. He not only despised our good preacher, but threatened that one day, as he had done to Hippocrates and Galen, he would set Luther's and the Pope's heads right. He also said that none so far who had written about the Holy Scriptures had grasped their right meaning.

A theologian recalled: 'I had several religious and theological discussions with [him]. If there was a trace of orthodoxy I failed to notice it. Instead, he talked a lot of magic of his own invention.' Meanwhile his fame continued to spread, and his lectures attracted ever-larger crowds. His most gifted students hung on his every word: iatrochemistry was the way forward, and they regarded themselves as his disciples. Nothing could stop him now. At last he would achieve his due, and take his place as one of the great men of his age.

Amidst all this it's difficult to see how Paracelsus found time to perform his mundane duties as town medical officer. But find time he did, and the town physicians and apothecaries were soon rueing the day Paracelsus had turned up in Basel. Not content with lecturing on iatrochemistry, he determined to put it into practice. He insisted upon making a tour of inspection of all the local apothecary shops. With predictable results. The shelves of infusions, fumigants, unguents and balsams were condemned as worthless; the prescriptions the apothecaries prepared were dismissed as 'foul broths'. To the delight of his students, he declared, 'After my death my disciples will burst forth and drag you to the light, and shall expose your dirty drugs, wherewith up to this time you have encompassed the death of princes.' In his own 'kitchen', aided by the hapless Oporinus, he began mixing his own medicines and distributing them free of charge to the sick and the needy.

The physicians fared little better. Just because they had passed exams at Nuremberg, it didn't mean they had a licence to fleece their patients. Even their most rich and powerful patients may have been too overawed to question their practices, but Paracelsus was not. Their bleeding and torturing of the sick was nonsense. Their treatment of wounds by padding them with moss or dried dung was positively harmful. The pills and potions and lotions they prescribed were unnecessary. The physicians' aim should be to cure the patient, not to make themselves rich and siphon money into the pockets of their cronies the apothecaries.

But his advice wasn't all negative. The years spent treating all manner of patients and their ailments on his travels through the length and breadth of Europe had left him with considerable expertise in the practical treatment of illness. And his theories in this field drew on a wealth of experience – with only the occasional taint of perversity.

Even his notorious doctrine of signatures in fact echoed a much deeper understanding of good practice. Nature had its own powers, and these must be allowed to work their way unhindered. 'If you prevent infection, Nature will heal the wound all by herself.'

This may have been sound practice, but it left medicaments and potions on the apothecaries' shelves unsold. The robust simplicity of Paracelsus' approach also dispelled the expensive mystique of the physician's art. Previously Paracelsus had been merely an object of ridicule and outrage. Now things were becoming more serious. He was beginning to hit people's pockets. But Paracelsus' distribution of free medicine, and free medical advice, made him a popular figure amongst the poor. And in Frobenius and his humanist colleagues he had powerful friends. The man may have had his weaknesses, but his thinking was in accord with the new ideas that were sweeping Europe. Here was a new broom to sweep out all the old superstitions and prejudices of the past.

Then disaster struck. Just five months after Paracelsus had delivered his first lecture at the university, Frobenius died while on a trip to Frankfurt. Paracelsus' enemies seized their opportunity, spreading the rumour that Paracelsus had poisoned Frobenius with his new chemicals. In fact the contrary was the case. Paracelsus had strongly advised the ageing Frobenius against making the four-hundred-mile round trip on horseback to

Frankfurt. His constitution was too weak. Paracelsus' prognosis had proved correct: Frobenius had died of a stroke, rather than of chemicals.

But Paracelsus' days at Basel were now numbered. Things finally came to a head five months later, over a court case. Paracelsus had been called in to treat a rich canon, who had promised him the sensationally high fee of 100 guilders if he succeeded. Paracelsus rapidly cured him with his own medicine. Whereupon the canon paid him only 6 guilders, claiming that this was sufficient for the time and effort he had expended on the case. Paracelsus took him to court. The local magistrates, amongst whom were several humanists, would set things aright.

But the tide had turned against Paracelsus, and the verdict went against him. This time it was Paracelsus' turn to be affronted. In a characteristic outburst, he publicly denounced the magistrates as corrupt and conniving. Libelling the city magistrates was a serious offence – punishable by a lengthy prison sentence or even the death penalty. A warrant was issued for Paracelsus' arrest. Fortunately he was tipped off, and managed to flee the city under cover of darkness.

Paracelsus' ten months' stay in Basel was the zenith of his career. Once again he was back on the road – sometimes passing a few comfortable months in the home of an admirer, at other times reduced to little better than a wandering worker of 'miracle cures'. One report describes him as being dressed in 'beggar's garb'.

In 1528 Paracelsus arrived at the flourishing commercial centre of Nuremberg. His reputation had preceded him, and he was quickly reported to the authorities as an impostor. In order to refute this charge, Paracelsus is said to have demonstrated his talents by curing some cases of elephantiasis which had defeated all the local physicians. If he did in fact succeed in this cure, it was an extraordinary achievement – for no cure for this complaint was found until the late nineteenth century. Yet according to his biographer Hartmann, testimonials proclaiming Paracelsus' feat 'may still be found in the archives of the city'.

Two years later Paracelsus once again visited Nuremberg. (Or perhaps this was the same visit – it is difficult to pinpoint his constant journeying. There is also a resemblance to the previous story, though once again there is evidence attesting to its veracity.) This time Paracelsus is said to have upset the authorities by publicly ridiculing the teachings of Galen and Avicenna, which still formed the bulk of the syllabus at the prestigious local medical college. As a result, Paracelsus was asked to demonstrate the superiority of his new method by curing some patients of syphilis, which was sweeping through the continent during the early decades of the sixteenth century. (It is generally believed that this disease was brought to Europe from the newly discovered Americas by Spanish sailors, though some commentators now believe that the leprosy mentioned in the Bible was in some cases syphilis.)

In these early days of the disease (or of its virulent resurgence) its manifestations were horrific and agonizingly painful. The patient's skin was covered in pustules, which developed into gaping sores, while the flesh rotted from the bones amidst a stench of suppuration. The fourteen syphilitics of Nuremberg had been confined in quarantine in a stockade outside the city walls, the local physicians having despaired of curing them. Paracelsus is said to have cured nine of these outcasts. And once again there is evidence of his feat in the city records.

In 1530 Paracelsus published a comprehensive description of syphilis, the first truly clinical outline of this disease to appear. He also claimed that it could be cured if the patient was given strictly limited dosages of mercury compounds, taken internally at prescribed intervals. It is now known that other physicians had begun using this treatment some years previously. Whether Paracelsus knew of this is uncertain. Mercury compounds were to remain the standard treatment for syphilis; these caused almost as much pain as the disease, as well as being so poisonous that they sometimes succeeded in killing off the patient. It was this treatment which inspired the popular saying: 'a night with Venus leads to a life with Mercury'. This remained standard practice until the revolutionary Salvarsan treatment of 1909 substituted arsenic compounds.

Four years later Paracelsus was applying his homoeopathic-cum-inoculation method to victims of the plague at Stertzing. Here he is said to have cured patients (and/or inoculated them) by administering pills of rolled bread impregnated with a tiny amount of infected patients' faeces.

Now in his early forties, Paracelsus decided it was time he wrote down his vast medicinal knowledge in some systematic form. He eventually published this in 1536 as *Die Grosse Wundartzney* ('The Great Surgery Book'). This work was soon in great demand, and for the first time Paracelsus found himself comparatively well off. He was now so famous that even princes summoned him for consultations. Yet his notorious and provocative behaviour ensured that he also remained an infamous figure. Even his faithful disciples found him difficult: 'He mumbles to himself for hours. When he speaks to others they can hardly understand him. He spends much time before his oven brewing powders but he does not suffer anyone to help him. He gets very angry when spoken to. Suddenly he has a fit and yells like a wounded animal. He gets impatient with the slightest mistake of his amanuensis.'

Yet the quality of his work was unsurpassed. Take his description of 'diseases of tartar' – by which he meant gout, arthritis, gallstones and similar ailments. These were all prevalent in medieval times, owing to the rich but unbalanced diet. Few escaped painful gout in old age.

According to Hippocrates, gout was caused by an imbalance

of the four humours, which resulted in a 'defluxion' restricting the flow of humours into the foot. For Hippocrates, gout, arthritis and similar diseases were manifestations of the ageing process, and thus incurable. Paracelsus disagreed. These were chemical diseases, which could be treated by iatrochemistry. They resulted from a build-up of tartar – which in arthritis seized up the joints and in gout deposited painful nodules of tartar in the foot. This was caused by the same process which deposited tartar in wine barrels, and on the teeth. Tartar in the body came from food and liquids, and was usually dispelled by the digestive process. In some cases the digestion was faulty, or the local water had an excess of tartar. (Paracelsus boasted that in his native Switzerland the water was so pure that no one suffered from gout, arthritis or gallstones.) If the disease was not too advanced, it was possible to expel the tartar by ingesting a substance that reacted with it, such as Rochelle salt (potassium sodium tartrate). Here in compelling detail was an early description of how chemical imbalance could cause a disease: one of the first instances of a genuine scientific medical aetiology.

One of Paracelsus' most effective medicines was laudanum. This was a concoction of raw opium, which he used to alleviate a wide variety of complaints. Indeed there were times when he seems to have regarded this almost as a panacea. He claimed to have invented this medicine, but he probably brought it back from his travels to Constantinople. At any rate, he certainly named it, perhaps from the Latin *laudare*, to praise. He also kept the secret of its ingredients to himself for many years. When wealthy patients demanded to be treated with this miraculously soothing new medicine, Paracelsus would charge them a fortune, claiming that it contained leaf gold and unperforated pearls. According to one contemporary description, he administered laudanum as pills 'which had the form of mouse excrements'. Later the name would be used to describe a solution of opium in alcohol. Paracelsus' laudanum remained a major weapon in the medical arsenal until late in the nineteenth century, its over-zealous use often resulting in addiction. A favourite tipple of the Romantic poets, this was the 'opium' to which de Quincey, Baudelaire and Coleridge were addicted. (Paracelsus was far from the last to regard narcotics as a panacea; over three hundred years later the young doctor Freud in Vienna would make the same mistake with cocaine.)

Paracelsus was equally cavalier in his prescriptions of chemical remedies. Over-confidence in his infallibility frequently caused him to dose his patients with mercury compounds and antimony, despite the fact that these substances were known to be toxic. The worthy physicians and apothecaries who opposed Paracelsus' methods may not have been motivated entirely by ignorant superstition and selfish commercial considerations.

In Paracelsus we can discern the elements of chemistry beginning to emerge through the noxious smokescreen of alchemy. But what of the elements themselves? Here, too, Paracelsus excelled himself. Zinc ore had been known since prehistoric times, and so had brass, an alloy of copper and zinc. However, Paracelsus was the probably first to realize that zinc was metallic. He also discovered a method for isolating metallic arsenic, mixing the sulphide with eggshells and describing the resultant metal as 'white like silver' – although Albertus Magnus probably preceded him in the actual isolation of this element.

Paracelsus may well have been the first to describe the properties of two other elements – bismuth and 'kobold' (cobalt), though he certainly wasn't their discoverer. Paracelsus first encountered bismuth in the Fugger mines. The Austrian miners of the period were firm believers in the Aristotelian theory of metallic evolution. According to this, the long process of natural transmutation resulted in three distinct types of lead: ordinary lead, tin and bismuth. The latter was closest to silver. As a result, when they uncovered a new vein of bismuth, the miners were in the habit of exclaiming: 'Alas, we have come too soon.' In another few years, they believed, it would have become silver.

Paracelsus described the properties of bismuth, but without of course realizing it was an element. He simply found that he was unable to break down this substance any further with the techniques at his disposal. Chemistry was not yet a coherent science, more a growing proliferation of techniques. But it was becoming evident that something was emerging from the alchemical hell's kitchens: a subject that was distinctly separate from the quest for gold. It has been claimed that Paracelsus was the first to refer to this subject as chemistry.

Paracelsus was also the first to mention the metal cobalt, or 'kobold'. Compounds of cobalt had been known since ancient times. They were used by the Egyptians and the Greeks to stain glass and artificial jewellery, giving it an alluring translucent blue colour; examples of this were to be found in the tomb of Tutankhamen. The word 'kobold' stems from the Greek word *kobalos*, the name given by superstitious ancient miners to the malicious gnomes found in subterranean mine shafts. These evil presences were said to cause rockfalls and explosions, and sometimes to bewitch miners. (Our word goblin has the same root.) All cobalt compounds were believed by miners through the centuries to be highly poisonous, and placed in the mines by the 'kobolds'. Goethe even refers to these demons in his *Faust*. Cobalt had first been isolated, albeit unwittingly, by alchemists in the Middle Ages, but Paracelsus seems to have been the first to recognize it as a new metallic element.

For the first time in around two millennia new elements were beginning to be found. The new techniques accounted for the discovery, around this time, of another metallic element: antimony. This too had been known since ancient times, but only in its sulphide compound. It was used by Middle Eastern women to darken their eyes and eyebrows, in order to increase their seductiveness. There are several references to this practice in the Bible, the best known involving the notorious Jezebel, who 'painted her eyes and adorned her hair, and stood looking down from a window'. (From which she was later thrown, and her dead body devoured by dogs.) The Arabic name for the substance with which Jezebel painted her eyes was kohl. By a series of misapprehensions this word came to be used to describe distilled liquids, and eventually the distilled liquid al-kohl – which became alcohol. The origins of the word antimony are even more far-fetched. They involve the legendary fifteenth-century abbot and alchemist Basil Valentinus (now known to have been the pseudonymn of one Johan Thölde, a respectable sixteenth-century German city councillor who practised alchemy but wished to keep his job). One day after work Valentinus is said to have emptied some crucibles containing antimony out of his cell window. This was eaten by pigs, who then became sick. When the pigs recovered they ate vast quantities to make up for their lost weight. But because they were pigs, and lived up to their name, they ate far too much, rapidly putting on excess weight. Valentinus seized upon this as an excellent way of fattening up the monastery pigs for Christmas. Then he decided to go one step further. As abbot, he felt the monks in his charge were also in need of a little fattening up for Christmas, so he covertly introduced some antimony into their diet. Unfortunately, many of the ascetic monks had bodies so weakened by fasting that they died before they could fatten themselves up. The substance they had eaten became known as 'anti-monakhos' (anti-monk, thence antimony). A likely story. Sadly, spoilsport modern commentators have pointed out that the name antimony was mentioned a few centuries prior to the legendary Valentinus, by Constantine of Africa in his translation of Avicenna's pharmacopoeia.

Despite Paracelsus' sterling work for chemistry, there is no getting away from the fact that he was also very much an alchemist in the old wizard tradition. Throughout his life he continued to search avidly for the philosophers' stone, which he felt certain was an elixir of life. At one stage he even suggested that he had found it, and partaken of it. He would live for ever, he used to tell his wide-eyed audience in the market square, before he was moved on to his next destination.

Yet even in his alchemy Paracelsus was capable of maintaining a highly scientific-chemical outlook. In his view, the universe had been created by a supreme chemist. He believed that the creation myth as related in the Bible was no more than a chemical allegory, describing a seven-day macro-experiment. (This may explain his claim that he alone understood the Holy Scriptures, as well as explaining why he kept quiet about the actual content of this unique understanding. The notion of God the Alchemist might not have squared with Vatican theology.)

Because the universe had been created by a chemist, it obeyed chemical laws. Chemistry consisted of unveiling the secrets of how the universe worked. Paracelsus saw both the universe and alchemical experiment as 'carrying to its end something that had not yet been completed'. He compared this to the process of digestion, or cooking. Unfortunately such embryo scientific perceptiveness was also marred by vast quantities of the usual metaphysical-allegorical-mystical-astrological nonsense – accompanied by typical Paracelsian delusions of grandeur. True alchemical knowledge could be imparted only by a magus, a magician-prophet possessed of supernatural powers. Such learning had been passed on from magus to magus since the beginning of time. And there was of course no doubt about who was the magus of this particular era. Market-place audiences and disciples alike held their breath in the presence of this wondrous being, who on occasion appeared almost as if possessed by the Devil. Not for nothing did Goethe base his *Faust* at least partly on the character of Paracelsus.

In 1538 Paracelsus returned home once more to Villach, only to discover that his father had died four years previously. The good citizens of Villach had respected Paracelsus' father, but they wanted nothing to do with his notorious son. Paracelsus was driven out, unable even to take over the house and property he had inherited. Homeless as ever, he continued to tramp from town to town, through Switzerland, Austria and Germany. 'I do not know where to wander now. I do not care either, as long as I help the sick.' Yet the old ways died hard, according to eye-witness reports: 'He even challenged an inn full of peasants to drink with him and drank them under the table, now and then putting his finger in his mouth like a swine.' The years of wandering and drinking, occasional despair and habitual poverty, to say nothing of his untiring self-proclamation and magniloquence, had taken their toll. He was in his mid-forties, but aged beyond his years.

Paracelsus had long suffered from divine self-delusions, and believed that he alone understood the Bible. But apart from this he paid little attention to conventional religion. His alchemy was his religion, and its incoherent metaphysics supplied his theology. Even so, there are hints that at some stage he may have undergone a conversion to some more orthodox belief. There is evidence of a spiritual deepening in his writings: 'The time of philosophy [i.e. both alchemy and science] has come to an end. The snow of my misery has thawed. The time of growing has ended. Summer is here and I do not know whence it came.' He came to accept his neglect and poverty as a sign. 'Blessed is he to whom God gave the gift of poverty.'

In 1540, at the age of forty-six, he finally succeeded in gaining employment at Salzburg. Now aged and decrepit, Paracelsus was taken into the service of the prince-archbishop, Duke Ernst of Bavaria, who despite his high clerical position was also something of a dabbler in the dark art. A year later Paracelsus was dead.

As was only to be expected of a magus who had imbibed of the elixir of life, the exact circumstances of his death remain something of a mystery. Paracelsus took up residence at the White Horse Inn at Salzburg. As usual he seems to have quickly antagonized the local apothecaries, physicians, academics and anyone else who crossed his intellectual path. But the duke remained his friend: a source of some local irritation. On the night of 21 September 1541 Paracelsus is said to have suffered a heavy fall on his way back to the White Horse Inn. Nothing unusual in this, one suspects. But precisely what happened in the narrow darkened street that night will never be known. It was rumoured that there had been a brawl with some thugs, who may have been hired by the local physicians to beat him up, or worse. Either way, Paracelsus died three days later on 24 September.

Less than two years later Copernicus published his work placing the sun at the centre of the planetary system, and the scientific revolution began.

5

Trial and Error

It was several years before the full significance of Copernicus' idea began to be realized; by this time it was being reinforced by a number of earlier scientific writings which had previously been ignored or largely forgotten.

The generally held picture of the medieval age as all but incapable of geniune scientific advance is not entirely accurate. As with most sweeping historical generalizations, there were exceptions. This was the age that gave us the wheelbarrow and the collective masterpieces of the Gothic cathedrals, and also produced the first scientific explanation of the rainbow. The latter was the work of the thirteenth-century monk Dietrich von Freiberg, about whom little is known except that he may have been a pupil of Albertus Magnus at Paris. Even his name has come down to us in a variety of versions, ranging from Theodorus Teutonicus de Vriberg to plain Vribergensus – and we don't even know which Freiberg in Germany he came from. But there is no uncertainty about his *De Iride* ('Concerning the Rainbow'). Here mathematics and scientific research were combined to produce the first great work on optics since Aristotle. (This was the very subject that would inspire the supreme minds of ensuing centuries to some of their finest thinking: Descartes and Kepler in the seventeenth century, Newton and Kant in the eighteenth and Gauss, 'the prince of mathematicians', in the nineteenth.)

In an age which had practically no notion of experiment beyond the realms of the alchemist's den, Dietrich von Freiberg conceived of an experiment whose originality and perception is breathtaking. How could he study what caused a rainbow? By studying in detail each droplet of rain which formed it. But how could he possibly grasp this elusive pot of gold? By studying a magnified water droplet suspended in the sky with the light of the sun striking it. To do this, Dietrich simply filled a spherical glass flask with water and traced the passage of the diffracted and reflected light as it passed through the flask to produce the rainbow effect. By means of this experiment he was able to produce theoretical explanations of how the rainbow produces its different colours, why it forms an arc, why the primary rainbow frequently has a second fainter upper arc, and why two people standing side by side do not in fact see the same rainbow. Such theoretical thinking, based on experiment, shines like a beacon in a world of scientific darkness.

A century later there appeared a scientific, philosophical and mathematical thinker whose ideas should have changed the world, but somehow didn't. Nicholas of Cusa was too far ahead of his time; his visionary scientific ideas would only later be confirmed. And his experimental technique looked ahead to the cleanliness, precision and attention to detail which characterize the modern chemical laboratory.

Nicholas of Cusa was born in 1401, the son of a moderately prosperous Rhineland fisherman. He quickly revealed an exceptional mind, but first came to public attention after conducting research into the Donation of Constantine. This was the fourth-century document in which the Emperor Constantine had allegedly ceded domination of the Byzantine and Roman Churches to the Pope. It was widely regarded as the final evidence for the Pope's claim to supremacy. Nicholas of Cusa showed that this document was in fact a forgery, dating from the eighth century.

At the age of thirty-six he was appointed to the delegation which set out from Rome for Constantinople to negotiate a reunion of the Byzantine and Roman Churches, a task which had previously defeated all comers, including Thomas Aquinas. But Nicholas of Cusa's delegation at last succeeded in paving the way to an agreement. (It wasn't his fault that this agreement lasted little more than a year. The Byzantines then decided that they hadn't really agreed to anything at all – thus ending all hope of assistance from the West and ensuring their fall to the Ottoman Turks fifteen years later.)

In 1440, at the late age of thirty-nine, Nicholas of Cusa was ordained as a priest. Yet he now began to publish major works expressing highly unorthodox ideas. The first was called *Idiota de Mente*, which could be translated as 'An Idiot Speaks His Mind'. However, in those days *idiota* stood for any layman or private individual holding no public office. This work consists of a dialogue between a philosopher, representing traditional Aristotelian views, and the aforementioned idiot. Interestingly, it is the idiot who puts forward Nicholas's views. These express a Platonic mathematical-scientific view of the world. 'The plurality of things arises from this, that the mind of God understands one thing in a certain way and a second in another way.' Mathematics is the mind of God, and this is the way the world works. 'Number is the principal clue which leads to wisdom.' The use of number leads to scientific discovery. Pythagoras had believed that the world was ultimately mathematical; Nicholas of Cusa introduced the idea that applied maths was the way to know the world. This brings practical knowledge. 'Mind alone counts; if mind is removed, distinct numbers do not exist.' Measurement, the matching of discrete parts of the world to numerical quantities – here lay the key.

The idiot who spoke these words was the salvation of Europe. This was the type of man who rescued the Western mind from stagnation. The layman who operated in the real world of commerce and practical work, yet also thought about philosophy –

only such a man could give birth to science. Nicholas of Cusa may have been a priest, and well versed in theology, but he understood that secular thought was the way forward.

At this stage China was far more advanced than Europe, yet from now on the torch-bearer of civilization would be the West. So what went wrong in China? The philosophers and the thinkers had become separated from the laymen and the merchants. The thinkers no longer combined with the doers. Nicholas of Cusa heralded this coming together in Europe: the union that would produce scientific thought.

Nicholas of Cusa believed that a learned man is one who is aware of his own ignorance. Such an attitude has its dangers. It can lead to resigned spiritualism (withdrawal from the world, mysticism, stoicism and the like), or to introversion (the 'know thyself', rather than the world, of Socrates). But in this case it didn't. Nicholas of Cusa saw it as a spur to further knowledge. For him it also alluded to the provisional nature of this knowledge, which was always open to improvement. 'As a polygon inscribed in a circle increases in number of sides but never becomes a circle, so the mind approximates to truth but never coincides with it . . . Thus knowledge is at best conjecture.' Once again the profoundly scientific notion of theory reappears, after an absence of around one and a half millennia.

Nicholas of Cusa was fond of using mathematical images to illustrate his philosophy. He even compares humanity's search for truth to squaring the circle. (This mathematical chestnut had come to obsess medieval mathematicians. The aim was to construct a square of equal area to a circle, using only a ruler and compass. Not until some time later was it finally proved that this is impossible. When a circle is drawn with a compass it will always have a radius of a measurable length, which can thus be taken as a unit of 1. So its area is $\pi \times 1^2 = \pi$. Therefore the edge of the square of identical area will be $\sqrt{\pi}$. But π is irrational: its value is 3.141592653589793 . . . and so on without repeating sequences of numbers for ever. In other words, it is by definition immeasurable. Therefore it is impossible to square the circle using a ruler and compass.)

Nicholas of Cusa's philosophical-scientific ideas may have been exceptionally advanced, but his actual scientific ideas were explosive. He believed that the earth revolved on its axis, and this led him to conclude that it moved around the sun. He also worked out that the stars were just like our sun, and they too must be circled by inhabited worlds. Further speculation led him to conclude that the universe was infinite. And since it had no central point, there was in space no such thing as 'up' or 'down'. Some of these ideas would remain ahead of their time until the dawn of the twentieth century.

Ironically Nicholas of Cusa reached many of these conclusions as a result of applying a principle of his own invention which was basically metaphysical. This was his *coincidentia oppositorium* – simply, the coming together of opposites. (Take, for instance, such opposites as a circle and a straight line. According to Nicholas of Cusa, these become coincident when extended to infinity: a circle with an infinite radius has a circumference which is a straight line.) He applied this principle both geometrically and theologically – often with the two intertwined. Nicholas of Cusa's ideas remained medieval in so far as they contained much theology. The significant factor is that in the *coincidentia oppositorium* he employed a new way of thinking to resolve previously insoluble problems. The *coincidentia oppositorium* was essentially a theoretical tool – one which had not been used by Aristotle, and so it enabled Nicholas to circumvent Aristotle's ideas.

Nicholas of Cusa's cosmological ideas were not based on experimental evidence, precise observations or mathematical calculations. Their theological guise, as well as the obscure original principle upon which they were based, may partly account for why Nicholas of Cusa didn't get into trouble with the Church. On the contrary: within ten years of becoming a priest he was made a cardinal, and later a bishop. Yet he didn't let such public appointments distract him from his intellectual pursuits.

Many of Nicholas of Cusa's ideas may have lacked experimental backing, but this was not through any lack of ability in the practical sphere. When he turned to practical matters he was without peer – his work ranging far and wide. Once again it was the notion of mathematically precise measurement which held the key to knowledge. In the practical sphere his main instrument was the scales. By measuring the different weights of a ball of wool, which absorbed water from the air, it was possible to determine humidity. By pouring equal weights of water into a square container and a circular container it was possible to calculate the value of π to a high degree of accuracy. But not all his experiments involved the scales. Like Roger Bacon before him, he suggested a much-needed modification of the calendar. (By this stage faulty calculation of the length of the year had caused the Julian calendar to drift almost a week out of kilter.) Nicholas of Cusa was a pioneer in the development of concave spectacles to correct short-sightedness; he suggested counting the pulse beat as a diagnostic technique; and he drew up one of the earliest reliable maps of Europe incorporating longitude and latitude. But the pinnacle of his practical skills was a piece of virtuoso scale work which was to have major implications for chemistry. This involved weighing a growing plant with great accuracy, day after day. His experiment was carried out with such precision he was able to discover that the plant was nourished by air, and also that air itself had weight. Here was something which no one before had realized. (Air, as one of the four elements, was regarded almost by definition as having no weight.) This was the first modern-style formal experiment carried out in biology. Its implications spread fissures through the accepted notions of physics, biology and chemistry. In time it would transform our understanding of what exactly reality is. However, such things could hardly be realized at the time – mainly because science had yet to develop the concepts with which to grasp such implications. It is this, more than anything, that probably accounts for why Nicholas of Cusa's ideas passed virtually unnoticed for so long. Even Copernicus had not heard of them when he worked out in mathematical detail the heliocentric plan of the solar system which Nicholas of Cusa had surmised.

Where Nicholas of Cusa had a hunch – as did several of the ancient Greeks – Copernicus produced a scientific model based on detailed observations, with mathematical backing. It is this, and the revolutionary effect of his theory, which ensures him his place in the scientific pantheon. Not without justification, his biographer Hermann Kersten claims that Copernicus caused 'the greatest intellectual revolution in the history of mankind'. Only Darwin, Newton or possibly Aristotle could be considered as contenders here.

Realizing the far-reaching implications of his theory, Copernicus withheld publication of his *De Revolutionibus Orbium Coelestium* ('On the Revolutions of the Celestial Spheres') until the very last moment. In May 1543, after suffering a stroke, the seventy-year-old Copernicus lay on his deathbed. According to his loyal friend Giese: 'the published version of his book was laid in his hands only in his last hour, on the day that he died'.

During the last years before Copernicus' death, rumours had begun to spread of his theory. Not only did it run counter to the Aristotelian orthodoxy of the Church, it was also dismissed by Luther, whose Reformation was now dividing the whole of central Europe. Fortunately, by May 1543 Copernicus was too far gone to notice that a preface had been inserted in his book by a Lutheran clergyman called Osiander, who had been left in charge of its publication. This preface was unsigned, and for years many people supposed it had been written by Copernicus himself. It stated that the theory contained in the book was not to be regarded as the truth. This was no description of the actual motion of the planets, merely a method which made it easier to calculate planetary motion for the accurate prediction of eclipses and the like. This had the effect which was probably intended. The book caused little controversy on its publication, and its few readers assumed that Copernicus himself didn't believe in the heliocentric theory. The book was perhaps purposely overpriced by the publisher, and was soon out of print. A second edition didn't appear until as late as 1566, and this was printed in Switzerland. It was almost certainly a copy of this edition which fell into the hands of Giordano Bruno, who was then studying in Naples. Every revolution needs its propagandist, and Bruno was to take on this role for the Copernican revolution. (Even though, as we shall see, his motives were far from clear, and his achievement was not exactly what he intended.)

Bruno was born in 1548 in the small town of Nola, some 20 km east of Naples. (Nola, which is in the Campania region, is according to legend the place where church bells were invented in the fifth century, hence the Italian *campana* and our word campanile for a tower containing bells.) Nola is also close to the dormant volcano Vesuvius. In later life Bruno recalled climbing Mt Vesuvius in his youth. From the distance its slopes had looked 'dark and drear against the sky', but as he had come closer he had seen that they were covered in flourishing vineyards, orange groves and olive trees. 'Astonished at this curious transformation, I realized for the first time that sight could deceive.' According to his account, he developed a questioning mind, and was soon asking himself: 'What are the grounds of certainty?' There's no denying his philosophical outlook, yet precisely how deep these doubts went remains problematic. As we shall see.

At seventeen Bruno was studying in Naples at the renowned

Dominican Convent. Despite its name, this institution conformed to the prejudice of its era: it would have been unthinkable for a woman to be admitted to a place of education. Bruno himself was soon contemplating other realms of the unthinkable, and established a reputation for unorthodox opinions bordering on the heretical. He read and absorbed the views of Erasmus (who was forbidden) and Paracelsus (who was ridiculed), and was not afraid of defending his views in passionate fashion. Despite his evident unsuitability, Bruno was ordained in 1572.

By now he had discovered his major inspiration: Nicholas of Cusa. Bruno seems to have read many of Nicholas' major works during this period. While still in Naples he adopted, and proclaimed, Nicholas' heliocentric view of the planetary system. This was confirmed by his reading of Copernicus. But Bruno developed this idea further. Like Nicholas of Cusa, he believed that each star was similar to our solar system, and that the universe contained a host of other inhabited worlds. He also declared that the universe was infinite. Bruno's cosmological views are almost identical to Nicholas of Cusa's, but his manner of expressing them reflects the change in world-view which was gradually taking place. Bruno's cosmology is not overtly interwoven with theological considerations. By now the ideas of the Renaissance had begun to take root. Knowledge could be expressed in rational humanist terms, echoing the classical era. The way was even opening up for a more scientific approach in science itself. All this makes Bruno appear more scientific.

Unfortunately Bruno's work is not all that it seems. He was ultimately scientific neither by temperament nor by belief. His Neapolitan tempestuousness could be overcome in moments of repose when he wrote down his ideas; but for the most part his deeper beliefs remained concealed beneath this scientific veneer. For centuries Bruno was regarded as a great propagandist for the scientific revolution. History is what it appears to be, and the appearance became the man. Only recently has it come to light that in truth Bruno had a very different agenda from the scientific one he dissimulated and disseminated with such far-reaching effect.

Some time while he was at Naples, Bruno came across the works of Hermes Trismegistos, the legendary 'Egyptian' alchemist. These converted him to a world-view which went far beyond that of the Renaissance. The Renaissance saw itself as reviving the classical learning of the ancient world. Hermes Trismegistos expressed an even earlier knowledge, the original knowledge – that which had inspired the classical writers themselves, and which had given birth to ancient learning. Hermes Trismegistos spoke of ancient Egyptian knowledge. This was the *prisca theologia*, the pristine (or pure) theology which had inspired all others. In the writings of Hermes Trismegistos could be found the leading ideas which had later surfaced in Pythagoras and Plato, ideas which found their echo in the teachings of Christ. Even the Church reluctantly acknowledged certain aspects of this. Hence Hermes Trismegistos was honoured as a gentile prophet – a similar status to that conferred upon Plato and Aristotle, who had both contributed so much to Christian theology but couldn't be regarded as Christians for obvious historical reasons. However, Bruno impetuously went one further than this. Secretly he believed that the Renaissance had only partly begun. The true Renaissance was yet to come. This would see the rebirth of the *prisca theologia*, where Christianity would be surpassed by the original true religion of history, the pure theology of ancient Egypt, as featured in the writings of the mythic magus Hermes Trismegistos.

For those so inclined, the writings of Hermes Trismegistos made a compelling case for such expectations. And this was no abstruse interpretation. It was all there in the text – Pythagorean numerology, Platonic ideas, Christian beliefs. There was only one snag: what we now know, and Bruno apparently did not, is that the writings of Hermes Trismegistos did not date from ancient Egypt, as they purported. They had in fact been put together in Roman times, when Neoplatonic and early Christian ideas were very much a part of the metaphysical scene. It was thus not surprising that this hermetic brew of alchemy, mysticism and whatever should also include these notions.

In the light of Bruno's secret beliefs, it is astonishing that his views appeared, and still appear, so scientific. Yet there's no denying that they do. And this is how they struck his contemporaries: including those who regarded themselves as his superiors – especially in his ordained vocation. But Bruno was temperamentally averse to conceding superiority in anything, especially in intellectual matters. He wouldn't be dissuaded from his 'scientific heresies' by mere Aristotelians, and made little effort to hide this fact from his superiors at the Dominican Convent. By 1576 they'd had enough. Wheels were set in motion to try Bruno for heresy – a charge which seldom failed, and frequently led to burning at the stake. Fortunately proceedings moved at a Neapolitan pace, allowing Bruno to disappear before things got serious.

As we can see from the case of Bruno and Hermes Trismegistos, chemistry – in the form of alchemy – was still capable of casting its shadow over the scientific revolution. At this point chemistry was in the curious position of being both ahead of this revolution and behind it. Chemistry's reliance upon practical experimental work pointed the way ahead, yet the theoretical beliefs which informed it harked back to the shaman and the witch doctor. As such, it could be said that chemistry represented humanity complete, in all its stages of development. This would be a supreme achievement in any field – except science.

The scientific revolution which followed Copernicus was mainly in the field of physics – with the initial breakthrough in astronomy leading to a wealth of technological and theoretical discoveries in other fields of physics. This revolution appeared to have nothing to do with alchemy. Yet Paracelsus' predilection for results, and his study of cause and effect in iatrochemistry, showed that chemistry (or alchemy) was not entirely separate. And now, in the most preposterous way, Bruno also demonstrated how the two could be related. Science, even as a screen for the alchemical beliefs of Hermes Trismegistos, was still science.

But surely the involvement between these two sciences in Bruno's thought was purely contingent? Superficially, it would seem so. But such a view is more difficult to maintain in the light of what was to follow. As we shall see, alchemy (not just embryo chemistry, but full-blown sorcery) would continue to play a curiously anomalous role. It would remain there in the background throughout the entire course of the great scientific revolution which began with Copernicus and ended with Newton. Indeed, Hermes Trismegistos himself features in the writings of both these supreme scientists. Copernicus refers to him in what is virtually a hymn to the sun at the start of his *De Revolutionibus*. This passage deserves ample quotation, not least because it shows how science is capable of inspiring poetic, philosophical and even religious emotions in the practitioner, where these may well have no place in the actual science:

> But in the midst of all dwells the sun. For who, in this most beautiful temple, could place this lamp in another or better place than that from which it can at the same time illuminate the whole? Which some not unsuitably call the light of the universe, the soul or the ruler. Trismegistus calls it the visible God, the Electra of Sophocles the all-seeing. So indeed the sun, sitting on the royal throne, steers the revolving family of stars.

Such lyricism may be forgiven for touching base with unscientific origins, yet their mere mention in these terms gives rise to speculation. (What did Copernicus really believe?) Newton, on the other hand, was less circumspect – on occasion even quoting Hermes Trismegistos directly in his notebooks. Here Newton's Hermes Trismegistos (and by intention Newton himself) is referring to alchemy, not science: 'Yet I had this art and science by the sole inspiration of God, who has vouchsafed to reveal it to his servant. Who gives those that know how to use reason the means of knowing the truth, but is never the cause that any man follows error & falsehood.' Faith, metaphysical 'art', reason – a murky current, ever-present beneath the smooth waters of the scientific revolution. Even today we remain basically cave-dwellers inhabiting modern cities, according to the reductionists. If such is physiologically the case (admittedly somewhat exaggerated) – what of the mind? Age-old assumptions and beliefs are not always shed the moment it becomes apparent that they have become redundant. Our mind, our language, our ideas, even our inspiration – these are all prompted by the past, the previous, the apparently discarded. Such things seem to play a left-handed role in the very thought which has overcome them. A prime example of this remains the role of alchemy during the ensuing century.

Meanwhile, the prophet of the scientific revolution fled north to Rome, having eluded the authorities of the Dominican Convent in Naples. (In an all-too-imaginable scene of sanctimonious horror, Bruno's secret copy of Erasmus was found hidden in the convent privy.)

Rome in 1576 was no place for an independently minded thinker, especially one with the volatile southern temperament of Bruno. In its effort to combat the Reformation the Roman Catholic Church had instigated the Counter-Reformation. The teachings of the Church on all matters were considered sacrosanct and unquestionable – including Aristotle's geocentric system of the planets, and his four elements. The Inquisition was reintroduced in Italy and northern Europe to weed out Protestants and heretics.

Not surprisingly, Bruno was soon in serious trouble once more. The details remain mysterious. Bruno was apparently denounced to the Roman Inquisition as a heretic, and excommunication proceedings were begun. Then the body of the man who had informed on Bruno was found floating in the Tiber. Whatever happened, Bruno decided it was best to absent himself, in haste and secrecy. This time he even shed his monastic habit – thus virtually excommunicating himself.

Bruno now began several years of wandering. He seems to have supported himself in a variety of ways, including giving private tuition on a mental system which he had perfected for increasing the power of memory. In essence this mnemonic system involved 'placing' each memory on a large imaginary wheel. This wheel could then be turned in order to retrieve the placed memory. But this was only the start. Inside the first large imaginary wheel you placed five more concentric wheels. These too were used for placing memories. By revolving these wheels inside each other, it was possible to form combinations of memories so as to generate new knowledge.

So far, so good (if a little mind-bending). But here too all was not what it seemed. Behind the successful systematic technique lay a hidden metaphysical structure. Bruno's six wheels were in fact derived from the teachings of the fourteenth-century Spanish mystic and (allegedly) successful alchemist Raymond Lully. In the writings of Lully these six wheels had formed a mystico-logical system for combining all knowledge – so that in the end its user could understand everything in the universe in all its combinations. Likewise it also had echoes in alchemical practices leading to transmutation into gold.

Once again Bruno stands at the pivotal point. Mysticism and

alchemy led him to develop a successful and practical mnemonic system. A century later, the details of this system would be studied by Leibniz, the German rationalist philosopher – inspiring him to construct one of the first calculating machines, consisting of a system of concentric wheels. In an echo of Lully's belief in the universal application of his system, Leibniz would be convinced that one day it would be possible to construct a calculating machine capable of solving all mathematical and logical problems. It would even be able to settle moral disputes: both sides would simply feed in their argument, and the machine would regurgitate the right answer. From mystic alchemy to a memory system to a calculating machine and the first inklings of the modern computer – each step accompanied by (and inspired by) its own more or less hidden misapprehensions. These continue to this day. We see modern computers as morally neutral, yet illogically retain the lurking fear that one day they might control the world. The idea that the computer can control the world is, in its different facets, both our fear and our inspiration – it is not difficult to see in this an echo of Bruno's alchemical inspiration for his memory system, and Leibniz' ethical delusions with regard to his calculating machine.

But Bruno's memory system was to have further, even more extensive ramifications. Besides being allied to Lully's alchemical system, it was also closely connected to a similarly far-reaching method of thinking which Bruno himself developed. This method was to play a major role in his thought, and marks the inception of a significant development in European thinking.

Bruno saw his new method of systematic thought as a form of creative logic: a way of thinking which would generate new knowledge. This had its origins in Nicholas of Cusa's *coincidentia oppositorium* – the method of viewing things by which opposites eventually came together. But Bruno developed this a significant step further. 'Profound magic is to draw the contrary out after having discovered the point of union.' The two opposites combine in the 'point of union', and out of this is drawn its opposite, 'the contrary'.

Just over two centuries later the German philosopher Hegel would discern in this statement the seeds of his own great dialectical method. As with Bruno's method, Hegel's dialectic involved two opposites coming together to generate something else. The two opposites in Hegel's dialectical system become the 'thesis' and the 'antithesis', and their 'point of union' (Bruno) becomes for Hegel the 'synthesis'. Bruno then 'draws the contrary' from this 'point of union'. Similarly, Hegel's synthesis itself becomes a new thesis, which then generates its own antithesis. These then combine to form a new synthesis, and so on. Hegel's dialectical method developed into a vast, all-embracing system, which accounted for God, the universe and everything else. Once again, Bruno's contribution had been pivotal. What for Nicholas of Cusa had been a mystical principle, Bruno developed as a dynamic method of thought. Hegel would then expand this into a vast interlocking metaphysical system (echoes of Raymond Lully's wheels within wheels). In a final elaboration, Hegel's dialectic would eventually be developed by Marx into dialectical materialism, a misplaced attempt to render this essentially unscientific method scientific.

In 1579 Bruno arrived in Geneva, a centre of Calvinism. Here he converted to the Protestant faith. But the Protestants proved no more tolerant than the Catholics. In their ideological war with Rome, they too had begun to lay down rigid doctrinal principles. They sought to return to the fundamental tenets of Christianity – abandoning what they saw as the fripperies and corruption of the Catholic Church. Ironically this conservative, back-to-basics approach meant that on scientific matters their doctrine was identical to that of the Catholics. Aristotle's geocentric universe and the four basic elements were still regarded as amongst the foundations of natural philosophy. When Bruno began preaching the Copernican revolution he soon clashed with a Calvinist professor over Aristotle. This time he was arrested and flung into jail. Once again, things could have turned nasty, but Bruno wisely chose to recant his views. As a result he was merely excommunicated and banished from the city.

Bruno had now succeeded in getting himself excommunicated by both sides of the Christian divide. This should have caused a God-fearing man such as Bruno some tribulation for fear of his soul. But Bruno was above such things. Soon would come the time when these two Christian opposites recombined. And out of this 'point of union' would arise the 'contrary': the *prisca theologia* of Hermes Trismegistos, the true and original religion of ancient Egypt. Fortunately Bruno had the sense to keep this to himself. For the most part, at any rate. From now on, apart from his notoriety as a propagandist for the likes of Nicholas of Cusa and Copernicus, we begin to hear the occasional rumour of his occult beliefs, accompanied by speculation that he was a magus.

Bruno's wanderings now took him to Toulouse, which was at the time a fanatically intolerant Catholic stronghold. Two reasons may account for this foolhardy move: Raymond Lully had once taught here and, unusually, the university had no religious requirement for its lecturers. Bruno may have had a Latin temper, but he also possessed Latin charm. After an interview in which he demonstrated his considerable intellectual skills and learning, the authorities gave Bruno a post as a lecturer in philosophy. In an uncharacteristic display of tact, he avoided natural philosophy. His lectures remained strictly orthodox, concentrating on Aristotle's *De Anima* ('Concerning the Soul'). In this work Aristotle declares that the relation between the soul and the body is an unnatural union. He compares it to the torture inflicted by the Tyrrhenian pirates, who would bind their captives to a corpse, a suitable metaphor for contemporary philosophy, which remained bound to the moribund corpus of Aristotelianism. By the late sixteenth century the renaissance in the arts was complete, the renaissance in science was just beginning, but the renaissance in philosophy still lay half a century in the future.

Curiously, at the same time as Bruno was lecturing in Toulouse the Portuguese philosopher Francisco Sanches was in residence and writing *Quod Nihil Scitur* ('Why Nothing Can be Known'). This all but forgotten masterpiece of profound philosophical scepticism argues that we can never really know the truth. It is possible to doubt our knowledge about anything. This very same idea was to be the starting point for Descartes – the thinker who sparked the seventeenth-century philosophical renaissance. Nicholas of Cusa's identity of opposites, Bruno's pre-dialectical system of thought, Sanches' methodic doubt – new ways of thinking were beginning to emerge, attempts to break out of the stranglehold of Aristotelian logic and doctrine. Bruno and Sanches must have met, but they would have been opposites in both temperament and thought: the quiet sceptic and the headstrong propagandist. Doubt and science – the time was not ripe. As we shall see, these opposites would eventually come together in the philosophy of Descartes.

In 1581 Bruno turned up in Paris, where the fame of his memory system came to the attention of Henri III. Regardless of Bruno's unorthodox views, he was appointed to the Court, which remained liberal despite the growing tensions of Catholic–Protestant conflict in France. Two years later Bruno travelled to London, where he was attached to the French embassy. Here he appears to have acted as a sort of low-level spy (and may also have accepted the odd payment as a double-agent for the English). In safely Protestant Elizabethan England, Bruno felt able to deliver himself freely of his anti-Aristotelian views. He travelled to Oxford, where he reduced an assembly of dons to near-apoplexy with his scornful dismissal of their antiquated notions. Had they no idea a scientific revolution was taking place? He also felt able to publish a number of occultist works, putting forward less scientific views.

In Bruno, occult beliefs and genuine science could be entirely separate. But, like his system of thought, these two opposites could also have a 'point of union'. The most notorious example of this is his attitude towards the Copernican planetary system. Bruno undeniably accepted the scientific truth of this system, but at the same time he also believed that it was a mystic-magical symbol of the cosmos. How Bruno's scientific thought could coexist with this metaphysical nonsense remains almost as much a mystery as the nonsense itself. But coexist it did. When Bruno talked plain science he talked plain sense. And as a result of his travels through Europe many became convinced of the truth of what he was saying. Fortunately, most of these converts were not aware of his esoteric beliefs, or simply chose to regard them as a personal eccentricity.

Scientifically, Bruno had found his way out of the Aristotelian quagmire. As in thought, so in life. Instead of medieval asceticism Bruno preached Renaissance humanism. We should be open to the world, not deny it. Bruno never married, but contemporary reports strongly suggest that he did not remain celibate. His detractor Mocenigo claimed: 'He told me that ladies pleased him well, but he had not yet reached Solomon's number.' It is safe to assume that Bruno did not rival the biblical Solomon with his seven hundred wives and three hundred concubines, but he seems to have enjoyed wine and women, if not plainsong. As in life, so in thought. Instead of a limited world beneath the dome of the revolving heavens, Bruno preached the solar system and an infinite universe of space filled with similar systems. And instead of the four Aristotelian elements, he preached atomism. How Bruno came by this idea is a brief history in itself.

As we have seen, the theory that ultimately matter consisted of indivisible atoms was originated by Leucippus, and developed by Democritus in Greece during the fifth century BC. Just over a century later it was taken up by the Athenian philosopher Epicurus, who gave us the word 'epicure' for one who believes in the sophisticated enjoyment of fine food and drink. Such a picture of refined hedonism is not an altogether fair reflection of Epicurus' philosophy. This taught a mechanistic world and believed in the restrained pursuit of happiness. The latter was achieved by withdrawing from politics and following the quiet life – which could be found through restraint of the appetites (especially those that the modern epicure prides himself upon).

Epicureanism remained popular for seven centuries, reaching its height during the glorious decline of the Roman Empire, when it did admittedly take on many of its more hedonistic aspects. With the advent of Christianity in Rome, austerity and repentance became the order of the day, causing the early Christians to identify Epicurus as the Antichrist, the figure of monstrous evil whose appearance would herald the end of the world.

Epicurus' moral philosophy may have been open to widely differing interpretations, but his natural philosophy was clear and scientific. Besides accepting a purely mechanistic world, Epicurus also adopted Democritus' atomic theory. The world was devoid of supernatural powers and consisted ultimately of tiny material particles which could not be created or destroyed.

Epicurus would not have been surprised that his philosophy outlived him by so long (though he might not have been so pleased at the form it took). Throughout his life he assiduously promoted his ideas, and is said to have written over three hundred treatises. Such is fate: only fragments of these survive. 'The idea of good is inconceivable if it does not include the pleasures of taste, of love, of hearing and sight . . . But virtue is no more than an empty word unless it means prudence in the pursuit of pleasure.'

Before Epicurus' work was reduced to fragments, it attracted the attention of the first-century-BC Roman poet Lucretius. The Romans added little to Greek thinking, though a number of their writers propagated Greek ideas. The finest of these was Lucretius, whose combination of poetry, scientific thought and anti-metaphysical philosophy was later dubbed 'almost as rare as the philosophers' stone'. Such exceptional qualities ensured that he was subjected to scurrilous slanders by early Christian commentators, for whom the scientific rational outlook was anathema. Thus we are forced to rely upon St Jerome for the few biographical details which have adhered to Lucretius' name. According to St Jerome, Lucretius was subject to periods of madness, brought on by a love potion which had been given to him by his wife. Only during his intervals of lucidity was he able to write poetry, which was 'edited' by his friend the great Roman orator Cicero. At the age of forty-four, during one of his fits of madness, Lucretius committed suicide. Insanity, sexual wantonness, the suggestion that someone else wrote his work, suicide – this sounds a bit like overkill. But saints are not meant to tell fibs; and poets, even philosophical ones, seldom live blameless lives – so perhaps even St Jerome's mountainous calumny may be based on a molehill of truth.

Lucretius believed that the universe had evolved, physically and biologically, and similarly that civilization was the result of sociological evolution. He was the first to divide history into separate ages of human development. He also had a rare disbelief in immortality. 'Death, that most terrifying of ills, is nothing to us . . . since when it comes we do not exist.' When we die our soul simply fades away 'like smoke'. 'In a short while the generations of living creatures are changed and like runners relay the torch of life.' The poetry is unmistakably Lucretian, yet how much the ideas are Lucretian or purely Epicurean is for the most part impossible to tell. Perhaps we should simply ponder

Lucretius' most famous remark: 'Nothing out of nothing comes.'

All these ideas, as well as a cornucopia of others, appear in his masterpiece *De Rerum Natura* ('On the Nature of Things'), whose very title is a conscious echo of Epicurus' lost work *On Nature*. The atomic idea is given pride of place in the first two books of this epic-length poem. Its prophetic accuracy speaks for itself. There are said to be an infinite number of atoms in the universe. These are of different types, but there are only a certain number of different types. These differ in weight, shape and size. They are all minute, solid and indivisible. However, they consist of inseparable parts, with the larger atoms having more of these parts.

This last, apparently contradictory, point was seen by many as a flaw in his atomic theory. If atoms have parts, they must ultimately be splittable. The advent of subatomic physics in the early twentieth century not only pointed to a resolution of this contradiction, but also indicated the miraculous prescience of this pre-science. What indeed was this early atomic theory? It had no possibility of experimental justification. The atomic theory was the result of pure theory (*theoria*: looking at, contemplation, speculation). Such astonishing insight could only have been coincidental. Yet as we shall see, whenever the atomic idea recurred in scientific thinking, as often as not it heralded a major breakthough. It is as if this idea possessed some talismanic quality for science.

Lucretius' *De Rerum Natura* achieved great renown in the Roman era, even inspiring the likes of Virgil, who was referring to Lucretius when he declared: 'Happy is the man who can read the causes of things.' Yet during the collapse of the Roman Empire Lucretius' great poem disappeared – the last known copies presumably going up in flames as the Visigoths sacked the eternal city, rampaging through the marble halls.

During the Middle Ages, the existence of the poem was known only through passing references to it in the works of others. Then in 1417 a sole surviving manuscript came to light. Half a century later *De Rerum Natura* was one of the first secular books printed on the new Gutenberg press, and it quickly became a bestseller. For a time Lucretius' poem became even more popular than Dante's *Divine Comedy*. It was read largely as literature, though its philosophy certainly appealed to the humanist mind. In contrast, its natural philosophy, i.e. the atomistic theory, was regarded as an anachronism – until it was read by Giordano Bruno.

After Nicholas of Cusa, Lucretius was to be the main influence on Bruno's scientific thought, although Lucretius and Nicholas of Cusa are opposites in all but their science. Lucretius dismissed all metaphysical thought; Nicholas of Cusa's thinking on all matters was imbued with theology. Yet Bruno managed to follow in the footsteps of both! For centuries he would be regarded as essentially a scientific humanist in the Lucretian mould, with just a Romantic whiff of occultism. Only later did it emerge that his scientific ideas were constructed on a template of dark occultism and metaphysics. But it should not be forgotten that it was the Lucretian element which he publicized. It was this which characterized the public face of both the man and his science – from the Copernican system to atomism. And it was this which was to be Bruno's downfall.

Leaving England before he was found out (as a spy, counter-spy, heretic or what-have-you), Bruno returned to Paris. But soon, 'Because of the tumults I left Paris and went to Germany.' His wanderings now took him as far afield as Marburg, Prague and Zurich, en route making friends, and in turn quarrelling, with many of the finest intellects of his day. These savants paid close attention to Bruno's ideas, but in the end were forced to dismiss the man who produced them. The Church, on the other hand, was beginning to take precisely the opposite point of view.

In 1591, while visiting the Frankfurt book fair, Bruno received an invitation from Venice to teach his memory system to a nobleman called Zuan Mocenigo. Bruno was forty-two, and had been away from Italy for twelve long years. He reckoned that by now it was probably safe for him to return.

Bruno travelled to Venice, making a side-trip to nearby Padua, where he had heard that the chair of mathematics was vacant. Perhaps he could obtain a genuine professorship, instead of delivering occasional courses of lectures and having to tutor ignorant noblemen. But he failed to get the job (which was taken the following year by Galileo).

In Venice, Bruno was put up at Mocenigo's mansion. From the word go, they didn't get on. Mocenigo was not overly bright. He became resentful and suspicious when he couldn't comprehend Bruno's system, and soon grew jealous of Bruno's reputation as it spread throughout Venice. After a series of increasingly heated disagreements, Mocenigo denounced Bruno to the Venetian Inquisition for holding heretical theories. Bruno was arrested and flung in the dungeons. (Just 150 years later Casanova would find himself in these same dungeons on a similar charge, though he soon effected a spectacular escape.) Bruno saw no need to try and escape. The court proceedings were fairly easy-going, and he felt well able to defend himself.

There is no surviving portrait of Bruno at any stage of his life. And it is only now, at his Venetian trial, that we encounter the first descriptions of his appearance. Appropriately, these seem incompatible. The clerk of the court described him as looking middle-aged, of medium build, with a chestnut-coloured beard. Meanwhile Ciotto, a local bookseller who was called as a witness, described him as short and thin, with a black beard. At his trial Bruno is said to have spoken rapidly, in the southern Italian manner, accompanying his words with vivid gestures and rapid changes of expression. It would seem probable that he always spoke like this. Others, in other contexts, described his earnestness and supreme concentration while delivering lectures, sometimes absent-mindedly perching on one leg as he did so. Such a demeanour gives the impression of a man who was utterly convinced of what he was saying. Though how he could have been convinced of what he was saying, now that we know what he *really* believed, is another matter.

In court Bruno was willing to concede that he had perhaps committed a few minor theological errors. But these were not serious, as they were due to his investigations into natural philosophy. He insisted that he had no disagreement whatsoever with the teachings of the Church. And, amazingly, he believed this.

Here at last we can begin to resolve the apparent inconsistencies of Bruno's mind. It was certainly possible for him to believe what he was saying – though it would have been most unwise for him to explain why. His 'minor theological errors' referred to his Copernicanism, atomistic beliefs and the like. These could not be serious theological errors because they were intimately related in symbolic form to the original ancient Egyptian religion, which he believed would soon transcend Christianity. And if this was the case he could have no quarrel with the teachings of Christ, because these too derived from the original 'true theology'. Fortunately Bruno kept quiet about these sensational beliefs which underpinned his scientific peccadilloes, and it began to look as if the Venetian Inquisition would take a lenient view. Then disaster struck.

The papal authorities had got wind of the proceedings, and Bruno was summoned to Rome to face the notorious Roman Inquisition. Here he was to be interrogated for seven years. To begin with, Bruno adopted his previous manner. But as his inquisitors became increasingly aggressive and insistent, Bruno became increasingly obstinate. Maybe he felt that he was doomed whatever he said, and decided he might as well stick to his minor scientific heresies (which covered a multitude of major ones). He appears to have despised the ignorance and bigotry of his inquisitors, who refused to listen when he tried to explain the incompatibility of Aristotelian doctrine with more recent findings in natural science. In the end, the exasperated inquisitors demanded that he recant everything. Nothing less than a complete and unconditional retraction of all his theories would be acceptable. This time it was Bruno's turn to become exasperated. He insisted that he had nothing to recant, and that he didn't even know what it was that he was expected to recant. Upon hearing of these words, Pope Clement VIII ordered Bruno to be handed over to the secular authorities, who were instructed to deal with him 'as mercifully as possible and without the shedding of his blood'. Sadly, these fine words were merely a hypocritical euphemism. They meant that Bruno was to be burnt at the stake as an unrepentant heretic.

On 17 February 1600, Bruno was led out into the Campo de' Fiori (Field of Flowers), his mouth bound and stuffed with a gag so that he couldn't address the crowd of onlookers. He was strapped to the stake amidst the pyre of kindling wood, which was then set alight, and he was roasted alive.

What was it they were so afraid Bruno might say? When they'd read him his sentence of death, he had gestured defiantly at the court, declaring: 'Perhaps your fear in passing judgement on me is greater than mine in receiving it.' At the height of his agony on the burning pyre, when offered a cross to kiss he had turned his head away abruptly. It seems he remained true to his deeper 'Egyptian' beliefs. But if he had spoken of these from the pyre, no one present would have understood him, let alone believed him. No, it seems that what the Church authorities most feared was that he would repeat his natural philosophical heresies. They were afraid of science.

'The sun does not move,' Leonardo da Vinci had written in code, in the margin of his notebook, around half a century earlier.

Nicholas of Cusa had known it and Copernicus had finally given this truth mathematical backing. It had nothing to do with religion.

But in search of its own intellectual backing, religion had taken on board philosophy. The methods used to validate the truths of natural philosophy were now used to prove theology. Thomas Aquinas in the thirteenth century produced no less than five proofs of the existence of God. Some of these are quasi-scientific. For example, the argument from First Cause: God as the ultimate beginning of the chain of cause and effect. But theology took on board more than the method of philosophy – it also took over its content. Along with logic came ethics, cosmology, natural philosophy. It wasn't Aristotle's fault that his philosophy became the holy writ. And even if Christianity had accepted Democritus' atoms, instead of Aristotle's four elements, the end result would have been the same. Science, like ethics, was set in stone.

Now it is arguable that where morals are concerned humanity has progressed little, or not at all, since the Bronze Age. The Homeric hero and Arnold Schwarzenegger's *Terminator* face similar moral dilemmas. But their weapons, as well as the medical care they can expect, are worlds apart. The scientific human condition, unlike the moral human condition, is not inherently static, and to maintain that it is only leads to absurdity. Even if the Church had taken on board Democritus' unsplittable atoms, it would eventually have found itself in the position of having to deny that Hiroshima had taken place.

Yet the science of Bruno's period was in just as much of a muddle as religion. Neither understood for certain what was really going on. The Pope pronounced that the planets circled the earth. Bruno saw the Copernican solar system as a metaphysical symbol. It would take time for the human mind to think itself clear of such misapprehensions. What was required was a new way of seeing the world.

6

The Elements of Science

In the most literal sense, an entirely new way of seeing the world was discovered in 1608. The invention of the telescope is usually credited to the Dutch lens-maker Hans Lippershey, who sold his lenses as spectacles. By the end of the sixteenth century this had become a boom industry. The spread of printing all over Europe had led to a widespread increase in reading, and the consequent increased demand for spectacles. The boom in lens-making in turn prompted the discovery of both the microscope and the telescope. By the early years of the seventeenth century minds were expanding with learning, and as an oblique result of this the world itself expanded, on the micro as well as the macro scale. All over Europe, many kinds of apparently unrelated changes were beginning to affect our vision of the world around us. With these new changes would come new questions. (Often, in fact, these were the old questions asked by the ancient Greeks. What is this world which we inhabit? How do these new wonders conform to the elements which we already know?)

Lippershey himself didn't actually discover the telescopic effect; this was the work of an anonymous apprentice. According to the story, this idle and somewhat bored young benefactor of mankind was one day playing around with the lenses he was meant to be polishing. He noticed that when he placed two lenses before his eyes, and adjusted the distance between them, he could form a magnified image of a church tower across the fields. Lippershey immediately realized the importance of this serendipitous perception, mounted the two lenses in a tube and named this invention the 'perspicillium' (meaning 'an instrument for looking through'.) The first perspicillium was then sold to the Dutch government as a military device. As with so many military secrets, then as now, word quickly spread far and wide to anyone who was interested in such information. Within the year it had reached the ears of Galileo in Padua.

It was Galileo who would first promulgate the elements of the new science. Galileo was ambitious, and always had an eye for the main chance. He had now been professor of mathematics at Padua for over fifteen years. Despite being only twenty-eight at the time of his appointment, Galileo had succeeded – where previously Bruno had failed – largely through persistence and self-advertisement, and by encouraging well-placed patrons to speak on his behalf. Back in 1592 the forty-four-year-old Bruno had been a figure of international stature, but he had merely made inquiries about the post, was better known in northern Europe and had no patrons in Italy other than the treacherous dimwit Mocenigo.

By the end of the sixteenth century, Padua was widely regarded as the finest university in Europe, attracting students from as far afield as Poland and England. (It was one of these students who passed on information about contemporary Italy to Shakespeare.) The moment Galileo heard about the new perspicillium, he characteristically realized two crucial points about this invention: firstly, no one had yet realized the full scientific potential of the idea; and, secondly, the perspicillium had great commercial potential.

The first perspicillia were only capable of up to threefold magnification. Within months Galileo had perfected an instrument capable of tenfold magnification. He then presented his new instrument as a gift to the city of Venice, which at that time ruled Padua. Galileo proudly explained how any fleet attempting to invade Venice could now be seen the moment it rose above the horizon, thus giving the city's defensive authorities vital extra hours in which to prepare for the attack.

But this was no altruistic act of generosity on Galileo's behalf. The grateful authorities immediately doubled Galileo's rather meagre salary and made him professor for life. Also, Galileo had discovered that cheap perspicillia were already being manufactured elsewhere in Italy, severely limiting the chances of any killing on the market. When these cheap perspicillia arrived in Venice, Galileo dismissed them as mere toys – distinguishing them from what he claimed as his own invention by christening this a 'telescope'. This comes from the Greek words 'in the distance' and 'to see' – though like the idea itself, this too was pinched by Galileo from someone else.

Galileo Galilei was an ebullient, red-bearded extrovert, whose unconventional character and obvious charms hid a rather more complex nature. He was frequently short of cash, owing to extravagance, family debts, and a demanding mother who insisted upon being supported in her native Florence. When Mama eventually deigned to visit Padua, she was horrified to find that her favourite professor of mathematics was living with a fiery back-street Venetian called Marina, who was almost fifteen years his junior and had already produced two sons by him. Galileo fled to nearby Venice to stay in the palace of his aristocratic pal Sagredo, while the two women in his life screeched at one another and eventually fell to tearing one another's hair out. Galileo was a witty and erudite companion to his noblemen friends, but at home he preferred to avoid the slangings and aggravations of outrageous domesticity. He would retire to his study for hours, sometimes days, on end, his mind immersed in science.

Galileo was perhaps the first to comprehend what the new science was really about. (This understanding would be applied almost exclusively to physics: how things worked. Only then could it spread to chemistry: the study of matter and the elements.) Indicatively, Galileo's understanding was grounded in an exceptional practical ability. His insights into what could be done, and how to do it, made him a superb inventor. Galileo's inventions ranged from the first thermometer to an instrument for measuring pulse rates, from a horse-operated water pump to a sector for calculating the trajectory of cannon balls. Unfortunately, owing to financial ineptitude and the lack of patent laws, or simply because his inventions were ahead of their time, Galileo's brilliant devices never quite achieved the huge financial success envisaged by their inventor. Yet these practical accomplishments gave him a deep theoretical insight into the workings of physics.

Around fifteen hundred years earlier a few isolated Greek thinkers, particularly Archimedes, had produced various unrelated mechanical facts and theorems – but there was no overall conception of mechanics as such. It took Galileo, who came up with the central notion of 'force', to show that here was an entire branch of unified theoretical and practical learning to be investigated. But why was the idea of force so important? To paraphrase an example from a seventeenth-century textbook: Imagine a man restraining a horse. The power required to do this could not be measured. Instead, tie the horse's rein to a sufficiently heavy rock. The immeasurable human power can now be read as a force, measurable in terms of the weight of the rock required to restrain the horse. All motion (or prevention of motion) was a result of a force at work, and could be measured. Here was an entirely new way of measuring the world, gaining insights into how it worked, and adapting such insights to human advantage.

Galileo became the pioneer explorer in this new field of learning, which he called *meccaniche* (or mechanics, from the ancient Greek for 'a contrivance or machine'.) But even here old habits died hard: typically, Galileo 'improved' on Archimedes' notion of momentum, claiming it as his own. Yet at the same time, his works indicate that he understood as many as three of the laws of motion, which Newton was not to formulate until over seventy years later. Partly because of Galileo's approach to science, and partly because of the way science saw itself at the time, he produced no actual definition of force; neither did he encapsulate his understanding of motion in the form of laws.

Yet these omissions appear trivial when compared with the major development, which Galileo expressed in the clearest possible manner. This is the key to his achievement, and yet it is so spectacularly simple that it now appears obvious to us. Galileo combined mathematics and physics. Until then these two fields of learning had been treated as largely separate.

This separation was already marked in the fourth century BC, when Plato's Academy placed its emphasis on abstract reality and 'pure' mathematics, while Aristotle's Lyceum concentrated on material reality, which was analysed by means of selection, comparison and classification. It might appear that Copernicus had forestalled Galileo's application of mathematics to physics, but this was not so. Copernicus had regarded the movements of the heavens as a purely mathematical problem. Mechanical notions such as weight, momentum and force did not enter into his calculations.

Only when Galileo combined mathematics and physics was it possible to conceive of the notion of measurable force. And with that modern science was born. Applying mathematical analysis to the problems of physics gave rise to experimental science in the modern sense. For the first time, practical events could be assessed, broken down into their component parts and measured, all in exact mathematical terms. Similar events could thus be compared – and when they matched, general laws could be formulated. Galileo called such tests *cimento*, Italian for 'ordeal'.

An experiment was a test, to see how (or if) a certain procedure worked. Our word experiment similarly derives from an Old French word meaning 'to put to trial'.

All this represented a categorical breakthrough. Or did it? Surely this was just what the alchemists had been doing for centuries? Indeed they had – and not all alchemical experiments had been entirely devoid of mathematics either. Most recipes for alchemical experiments included at least an indication of the 'measures' of ingredients required, along with detailed descriptions of the procedures to be followed. Up to this point, alchemy was undeniably an experimental science. The parting of the ways came with the results of these experiments. In the majority of cases, only one result was sought – gold. Having failed to achieve this, the experimenter seldom felt the need to record what had been achieved. And others, who did, tended to claim bogus, fanciful or metaphysical results. No science could be built on such imaginative foundations.

But it has been claimed that Galileo was not even the first to conduct modern-style experiments which combined mathematics and physics. A number of contemporaries, a few years prior to Galileo, had begun using similar experimental methods. This claim has more than a little justification. As we have already seen, the entire mindset of the medieval world was crumbling. The old certainties attributed to authority (namely Aristotle) were being seen as increasingly uncertain, prompting all manner of new ideas to replace them. The application of mathematics to reality was just one of these new ideas. Such notions were 'in the air', and many were thinking along similar lines. Modern science was being born all over Europe, the creation of various individuals thinking independently in a scientific fashion.

It is customary to call such thinking 'ahead of its time'. This does not do justice to precisely what was happening at the beginning of the seventeenth century in Europe. These individual scientific thinkers and experimenters, often working in isolation from one another, were not so much ahead of their time as creating an entirely new time. All over Europe, from Poland to southern Italy, a new mindset was gelling. An indication of this is that several important discoveries were made, all but simultaneously, by different individuals who could not possibly have known of each other's work, let alone resorted to plagiarism. Here indeed was a new development. Science didn't just advance as a result of great discoveries by great men. Just as important as these individual geniuses was the advent of a new way of thinking – which could lead several thinkers to the same discovery at once. (Whereas without some new way of thinking no categorical advance could possibly be made: the four elements had already been discovered, so there was no need to explore here any further.) Without this new thinking, all thought about such things as the elements was moribund. With it, scientific minds were soon making all manner of simultaneous new discoveries.

One example will suffice. Galileo completed his geometric sector for calculating the trajectory of projectiles (cannon balls) in 1597. Just a year later, an uncannily similar device was produced independently in London by the Elizabethan mathematician Thomas Hood, though this failed to rescue him from penury. Meanwhile the Dutch mathematician Dirk Borcouts, who corresponded with Descartes, was also working on his own bronze sector for calculating projectiles. (This can still be seen in the local museum.)

Then why was Galileo so important? It was as if a host of these different trends came together in his mind – which showed itself to be superior in both quality and reach. Galileo's application of mathematical analysis, his experiments, his conceptual originality (for example, the notion of force), his consummate technical skill, to say nothing of his strokes of genius – these were what set him apart from his contemporaries. Galileo was not always the first to arrive at an idea (even when he genuinely thought he was), but his was usually the finest mind to do so. And it showed in the results. Nothing illustrates this better than Galileo's use of the telescope.

The original crude perspicillium produced an image that was upside down, as well as having a magnification of less than three. By the time Galileo had finished, he had perfected a telescope capable of a magnification of over thirty which produced an image the right way up. And while others saw the telescope as a military device, Galileo understood the full potential of his 'improved invention'. He raised his telescope to the night sky, and immediately an entire new universe was revealed. (Once again, Galileo was not the first to do this. Already the Englishman Thomas Herriot was using a telescope to map the surface of the moon. This largely forgotten Elizabethan pioneer was a man of many parts. He crossed the Atlantic and undertook one of the first ever anthropological studies, an investigation of 'the naturall inhabitants' of Virginia. Together with Sir Walter Raleigh and the playwright Christopher Marlowe, he was involved in the 'School of Night'. His intellectual pursuits were equally far-reaching: after mapping the moon, he established himself as one of the leading astronomers in Europe; he was to invent a simplified notation which transformed algebra; and he became an enthusiast for 'drinking' tobacco smoke as a panacea. Herriot was typical of the 'amateur geniuses' now being thrown up by the seismic shift taking place in the European mind.)

Wherever freedom of thought prevailed – especially in England and Holland – intellectual advance and excellence soon became evident in fields which had been comparatively neglected during the Renaissance. The achievements of philosophy, literature, mathematics and physics now began to surpass those of painting, sculpture and architecture. After aesthetics, science: form was now being filled with content.

When Galileo trained his telescope on the moon, he was surprised to see that its surface had unmistakable mountains and valleys. Making ingenious use of the shadows cast by the mountains, he was even able to calculate their height. He then turned his telescope on the group of seven stars known as the Pleiades. (They were named by the ancient Greeks after the seven daughters of the god Atlas, who killed themselves with grief after the death of their sisters, who formed the nearby constellation, the Hyades.) When Galileo looked through his telescope he found that the seven Pleiades visible to the naked eye now became over forty stars.

But it was not until he began studying Venus that Galileo made his major discovery. It soon became clear to him that Venus had phases, similar to those exhibited by the moon. At certain times it was a crescent, at others a half-sphere, then it would become full. As with the moon, the light from Venus was obviously reflected from the sun – and these phases showed that it revolved around the sun. Here was incontrovertible observational evidence that Copernicus had been right about the solar system.

As the twentieth-century philosopher of science Paul Feyerabend pointed out, Galileo's telescopic observations were more adventurous than even Galileo realized. 'They not only increased knowledge; they changed its structure.' Between them, Nicholas of Cusa and Bruno had suggested that the universe was infinite, that stars were other solar systems and that there were other earths – but all this had been mere speculation. Even the most advanced minds tended to the belief that – even if Copernicus was right – the heavenly bodies were different from the earth, just as Aristotle had maintained. Galileo had provided proof that this was not the case. The moon was similar to the earth, Venus was similar to the moon, and there were countless stars which remained invisible to the naked eye. All these were huge solid bodies moving through the vastness of space. When Galileo trained his telescope on the night sky, the entire structure of the universe changed before his eyes. Things could never be the same again.

Galileo was the first truly original philosopher of science since Aristotle. Following Pythagoras, he believed that the world could be described in terms of mathematics, and that mathematics held the key to investigating the world. But he considered that only certain aspects of the world could be described in mathematical terms. These he called the 'primary qualities – shape and size, number, position, and motion'. All these qualities were objective, and were properties of bodies. For example, it was possible to measure the size, shape, speed, etc., of a cannon ball. But there were secondary qualities – such as taste, smell, colour and sound. These were not measurable, because they did not belong to the bodies themselves. These qualities existed only in the mind of the person observing the body, they were merely an effect of the body.

This distinction was crucial. Science could move forward with what was measurable. Other, apparently unmeasurable qualities would be disregarded as mere subjective phenomena. With hindsight, we can see that the primary qualities belong to physics. The secondary qualities belong more to chemistry. In order to define itself, to clarify its vision, science had to limit itself. In order to establish the elements of modern science, what could be thought about clearly had to be separated from what could not. Galileo restricted science to the question: 'What happens?' He ignored science's concomitant question: 'What is it?' Physics can operate without the latter question, but it is a central perception of chemistry. However, by this stage the vision of chemistry had become hopelessly blurred. Arguably, its only significant use was the manufacture of medicines. Any large theoretical advance was hamstrung by the theory of the four elements and the confusions of alchemy. Before chemistry could advance, men first had to understand what science was through physics.

Science now progressed into a colourless, odourless, tasteless, soundless world – a barren universe indeed. For any branch of knowledge to become a science, it seems that such drastic reduction is invariably necessary. In the contemporary era, as economics aspires to become a science, it has been forced to reduce the richness of human nature to *Homo economicus*. This limited species is defined purely by what it consumes, what it produces and its constant greed for more. It seems that only by reducing the human being to a mere digestive tract can economics hope to achieve its salvation as a science.

Such reductions of the world to a scheme of things can have serious consequences for our view of the human condition. (Are we ultimately nothing more than consumers? Just statistics in the flow chart of human existence?) The scientific reduction which began with Galileo was to prove highly offensive to the human psyche, and remains to this day at least in part unacceptable.

Compared with the barren, colourless world of the new science, the old medieval world was exceptionally rich. For a start, it had an overall meaning. The world was to be contemplated, its deeper significance pondered upon. There was a hidden spiritual and ethical agenda. The universe could be read like a work of literature: God's book. Metaphor and symbolism permeated its entire workings. But the new science was not literary criticism. Here the world had no pedagogic intent, it did not test souls or illustrate metaphysical beliefs. It had no apparent cultural baggage. The new scientific world was shallow and philistine. It sought simple truths, rather than 'the truth'. Here was a universe bereft of ultimate meaning.

This too was in part a renaissance of ancient Greek cosmological thinking. The very word 'cosmos' derives from the Greek word which merely means 'order'. Natural philosophy, as practised by the ancient Greeks, including Aristotle, sought order in the universe, not meaning. Ironically, it was this very absence of theological content which would make it acceptable to Christianity.

Hence, by Galileo's time the world had already been fully explained and interpreted – by the combination of quasi-Aristotelian natural philosophy and Christian theology. The Bible was the key to the scientific world. Here it was plainly stated that the stars had been created by God to provide illumination for humanity. Thus there couldn't possibly be stars which were invisible to the human eye, as Galileo claimed after looking through his telescope. Such stars would be superfluous, meaningless, absurd, epithets which could hardly be applied to God's works. And as for the Copernican heresy, which Bruno had preached . . . Galileo was summoned to Rome to explain himself.

The wrangling continued over a period of years. Not all in the Church were entirely unsympathetic to the new science; many felt that an accommodation would have to be reached sooner or later. Galileo suggested a way out. The Bible should not be regarded as literally true. It was in fact just an ancient historical document, intended for moral guidance. Its authors had never meant it to be a work of scientific fact. But the argument in Rome was not being conducted entirely on scientific grounds, or even religious ones. A political battle was taking place between rival factions. The power, as well as the soul, of the Church was at stake.

In 1632 the sixty-eight-year-old Galileo was once again summoned to Rome. This time he found himself facing the Inquisition – all too aware of the fact that Bruno had faced this same body just thirty-two years previously, and on the same charge. Galileo's ebullience and bravado had always been fuelled by deep inner uncertainties: he was not made of the stuff of martyrs. When interrogated about the Copernican heresy, he soon began wavering. Eventually he capitulated, before there was any resort to torture. (In the play *The Life of Galileo* by Bertolt Brecht, Galileo is led to the door of a dungeon by his inquisitor, who indicates the instruments of torture within. While this scene has no basis in fact, it is metaphorically accurate.) On his knees, Galileo was forced to swear that he 'abjured, cursed and detested' his new science. Copernicus had been wrong, the earth was the centre of the universe. Yet even as Galileo rose to his feet, he is said to have muttered beneath his breath '*Eppur si muove!*' ('But still it moves'). The ageing, ailing Galileo may have escaped burning at the stake, but he was nonetheless sentenced to life imprisonment – which in practice became house arrest at his home outside Florence.

The shock waves quickly spread throughout Europe. In Holland, the French philosopher René Descartes was putting the finishing touches to his *Treatise on the Universe*, in which he had independently come to many of the same conclusions as Galileo. Descartes' approach was different from Galileo's experimentalism. For the French philosopher, the prime tool in the quest for knowledge was reason. To achieve a clear scientific view of the world, nothing less than an entirely new method of thinking was required.

From an early age Descartes had decided to dedicate his life to the pursuit of thought. For this he required a solitary existence, undisturbed by the clamours of everyday life – to say nothing of the tumultuous historical events which were now unfolding in Europe. The collapse of the Holy Roman Empire into Protestant and Catholic states had fragmented Germany, leaving a power vacuum in the heart of the continent. Commercial, dynastic and religious rivalries exploded into the Thirty Years War (1618–48), which soon involved countries from Sweden and Russia to France and Spain. This cataclysm would eventually lay waste large tracts of central Europe from the Baltic to Bavaria. Germany was reduced to smoke-blackened cities and unsown fields haunted by carrion crows, its remnant population wandering the highways and byways at the mercy of robber bands.

Descartes' reaction to all this comes as something of a surprise. In pursuit of the quiet life, the finest rational mind in Europe decided that his best option was to join the army. In 1618, the very year the continent was plunged into war, Descartes signed up for the Prince of Orange's army in Holland. But Descartes had not only had a great rational mind, he was also very astute. He would join an army only when he knew it wasn't going to war; and as a gentleman volunteer officer in such an army, he knew that he would be left much to his own devices. The extent to which this was the case is best illustrated by the fact that Descartes maintained a strict lifelong habit of never rising before noon. His mornings were devoted to lying in bed thinking.

In 1620 Descartes found himself attached to the army of Maximilian, Duke of Bavaria. This was encamped in its winter quarters in the snow-covered Bavarian countryside. (As with hunting, everything had its season: in the seventeenth century no army worth its salt even considered fighting once the weather turned bad.) In Descartes' own words: 'Winter set in, and I found myself in a spot where there was no society of any interest. At the time I was unworried by any cares or passions, so I took to passing my time alone with my thoughts, sitting in a stove.' The last remark is not to be taken literally: Descartes probably meant the small room in a Bavarian house which contains a large tile-covered stove.

In these cosy circumstances, Descartes now undertook a mental exercise which was to revolutionize Western philosophy. He began subjecting his entire existence to the scrutiny of reason. How do I know anything about the world around me? By the use of my senses. But I can be deceived by my senses. A straight stick looks bent when it is dipped into water. How do I even know that I am awake, that the whole of reality is not a dream?

How can I tell that it is not a fabric of delusion woven by some malicious cunning demon simply to deceive me? By a process of persistent and comprehensive questioning it is possible to place in doubt the entire fabric of my existence and the world around me. Nothing remains certain. But in the midst of all this there is nevertheless one thing which does remain certain. No matter how deluded I may be in my thoughts about myself and the world, I still know that I am thinking. This alone proves to me my existence. In the most famous remark in philosophy, Descartes concludes: '*Cogito ergo sum*' – 'I think, therefore I am.'

Now that he had established this one ultimate certainty, Descartes proceeded to rebuild upon this foundation all that he had doubted. The world, the truths of mathematics, the snowbound Bavarian winter, the nature of his existence – all returned, tested by doubt, but more indubitable than ever now that they were built on such an indubitable foundation.

As a result of his profound meditations in the midst of the Bavarian winter, Descartes conceived the idea of a universal science. This would be a method of thinking which would be capable of understanding all human knowledge. Such a cognitive method would not only include all knowledge, but would also unite it. This system would be based on certainty alone. Free from all prejudices and unwarranted assumptions, it would start from basic principles, which would be self-evident, and it would build from these.

Descartes was a superb original mathematician. (Cartesian coordinates, for plotting objects in three-dimensional space, were his conception and are named after him.) So it is perhaps not surprising that Descartes' all-embracing scientific vision bears a strong resemblance to mathematics. Descartes sought rigidity, rationality and certainty for science. Compare this to Galileo, who placed his emphasis on experiment, to which he applied mathematics. Even Galileo's theoretical division into primary and secondary qualities was prompted by experimental considerations: what could be measured, and what resisted this experimental method? Where Galileo sought a method of experiment, Descartes sought a method of thought. What was possible – as distinct from what was certain. They were both attacking the same problem (scientific truth), but from opposing angles (practice/theory).

But what precisely was Descartes' new method of thought? He outlines this in his treatise *Rules for the Direction of the Mind*. The universal science can be discovered only by thinking in a certain manner. This consists of two basic rules of mental operation: intuition and deduction. The first he defines as 'the conception, without doubt, of an unclouded and attentive mind, which is formed by the light of reason alone'. Deduction was defined as 'necessary inference from other facts which are known for certain'. Descartes' celebrated method – which came to be known as Cartesian method – lay in the correct application of these two rules of thought.

Here was a logical method for scientific progress. It accounted for the theoretical standing of scientific knowledge, which was based upon 'the facts which are known for certain', i.e. facts which derive from observation and experiment.

Descartes now began addressing himself to the latter, applying his method to the world. The book which resulted from this was called *A Treatise on the Universe* (*Le Monde*). In this, and in subsequent works, Descartes tackles a wide range of scientific problems. In place of the muddled Aristotelian idea of motion as a kind of 'potentiality' which existed within a body, Descartes set out three clear laws establishing inertia, momentum and direction. He also investigated practical problems such as the refraction of light as it passes from air into water, deriving a principle which he then applied to the question of how rainbows are formed. (By the scintillating use of reason alone, Descartes reached the same answer as Dietrich von Freiburg with his humble experiments three centuries earlier.)

Descartes came to the conclusion that the world operated in a mechanistic fashion. Objects collided and rebounded; the human body and the heavenly bodies functioned like clockwork; once set in motion, the cogs of cause and effect ground on irreversibly. As with Galileo, Descartes asked how things work, rather than what they are. Descartes avoided the second question by regarding matter as ultimately 'consisting merely in its length, breadth and depth'. This is of course matter seen with the clear vision of physics, rather than the messy recipes of chemistry. Though, curiously, Descartes appears at one point to have believed in Aristotle's four elements: 'The primary mingling of these four compounds results in a mixture which can be called the fifth element.' This fifth element was matter itself. Descartes was convinced that any consideration of Aristotle's four elements would soon be seen as irrelevant: the physical properties and behaviour of matter were what mattered in a mechanical universe. Both Galileo and Descartes viewed the world as mechanical. Galileo conceived of the notion of force, and created mechanics. Descartes attempted to explain the entire world in terms of a 'mechanical philosophy'. Either way, it was inevitable that both of them would soon come to the conclusion that Copernicus had been right.

Immediately Descartes heard of Galileo's fate at the hands of the Roman Inquisition, he gathered up the scattered pages of his *Treatise on the Universe* and locked them away in a drawer. Like Galileo, he was convinced that the earth and the planets orbited the sun. It was just a matter of biding his time until the Church came round to the same conclusion. (Pope John Paul II was to make a posthumous apology to Galileo in 1997.)

Galileo sought to lay down experimental guidelines; Descartes sought to develop a mathematical-mechanical philosophy. Meanwhile one man was already producing a science of thought and practice which combined, and even superseded, these attempts. And it was this combination which was to point the way forward. The pioneer responsible for this was the Englishman Francis Bacon. A man of exceptional but ill-balanced talents, Bacon lived when the brilliance of Elizabethan England was passing into the shadier realms of the Jacobean era. This provided him with a suitably glorious but dangerous stage on which to play out his wayward destiny. (Such were Bacon's talents that some have believed it was he, rather than the comparatively uneducated Shakespeare, who produced the body of plays to which Shakespeare put his name. Through the centuries this theory has convinced thinkers of the stature of Freud and Disraeli.)

Francis Bacon was born in 1561, while his father Sir Nicholas Bacon was lord keeper of the great seal, an office of state equivalent to cabinet rank today. Sir Nicholas was an able and principled man who had risen from modest circumstances amidst the new Elizabethan meritocracy. His example was always deeply admired by his son Francis, despite the fact that he seemed resolved to follow in his father's footsteps without following his worthy example. Francis's mother was a determined and interfering woman of Puritan principles, who was continually concerned about her son's moral welfare (and with good reason). The contradictions in Francis Bacon, unapparent to begin with, were always there. A brilliant scholar, he abandoned Cambridge at the age of fifteen, declaring his distaste for the 'unfruitful' Aristotelianism which prevailed. There is a portrait of the young Bacon dating from this period, by the exquisite miniaturist Nicholas Hilliard, the first great English painter. This miniature depicts a tousle-haired youth of slightly uncertain hauteur, wearing an Elizabethan ruff. Hilliard was a man of wit and some arrogance himself, and appears to have empathized with his gifted young sitter. Around the oval portrait, Hilliard has inscribed in Latin the motto: 'If only I could paint his mind.'

Bacon entered the Elizabethan political scene filled with the aspirations encouraged by his promise. But these aspirations were to suffer a blow when his father died intestate, leaving Francis short of money. Bacon already had an extravagant temperament, and would remain dogged by lack of sufficient cash throughout his life, with disastrous consequences. Aspiration soon hardened into ambition.

The Elizabethan era was the first hour of greatness in England's history. An essentially provincial country at the periphery of Europe, England's previous role in international affairs had been largely as a harassment to the French. Elizabeth's father, Henry VIII, had wilfully severed ties with Catholic Rome, declaring himself head of a Protestant Church of England when the Pope refused to grant him a divorce (from the first of his six wives). A quarter of a century later the young Elizabeth ascended to the throne. Quick-witted, fluent in five languages, icily beautiful, yet with 'the style to win the hearts of the people', Elizabeth ushered in an era of national self-confidence. The Renaissance had already reached England, but under Elizabeth it bloomed – in a peculiarly English manner, which blended elements of humanism and medievalism (epitomized in Shakespeare's *Macbeth*, with its machiavellian Macbeth and medieval witches). The country flourished culturally, socially and economically as never before. Elizabeth's great navy, with sailors such as Raleigh and Drake, began carving out an overseas empire; playwrights of the calibre of Christopher Marlowe and Ben Jonson were overshadowed only by Shakespeare himself; and at the heart of it all was the splendid court of the Virgin Queen, who proved as astute in the handling of her political advisers as she was capricious in the handling of her favourites (often the same men).

By the time Francis Bacon became a significant player on the political scene, England had survived the coming of an armada representing the might of Catholic Spain; but the country was now beset by grumblings of discontent. The striking redhead who had ascended the throne thirty years previously had become a vain, ageing spinster with dyed hair, her face chalk-white with caked powder, her extravagant gowns dripping with pearls. Elizabeth's court was riddled with intrigue, and her young favourites were increasingly untrustworthy.

There was little room for honour or dignity in such a world, and Bacon soon learned to dispense with both. He read *The Prince* by Machiavelli, and was fatally attracted by the Italian's unprincipled opportunism masquerading as political philosophy. After inadvertently offending Elizabeth, Bacon ingratiated himself with the Earl of Essex, her current favourite. Fortunately he wasn't in on the plot when the disgruntled, hot-headed Essex led an insurrection against Elizabeth. Genuinely shocked by Essex's treason, Bacon saw no clash of interest when he drew up the judicial report which led to his erstwhile friend's beheading.

In 1603 Bacon successfully negotiated the transfer of power from Elizabeth to James I, a feat requiring cool nerve and absence of scruple in equal measure. With an obsequiousness that turned even the stomachs of his courtly and ambitious contemporaries, Bacon now wheedled his way into the favour of Sir George Villiers (later Duke of Buckingham), the new king's favourite. As Bacon himself drily put it: 'By indignities men come to dignities.' He was soon rising through the upper ranks of political appointments, with high office in view. By 1618 he had even surpassed his father, becoming lord chancellor, the highest legal post in the land. Now he had money to indulge his extravagant fancies – which nonetheless ensured that he overspent. His country mansion at St Albans, twenty miles north of London, became a byword. According to his friend, the biographer of the age, John Aubrey: 'When his lordship was at his house at Gorhambury,

St Albans seemed as if the court were there, so nobly did he live. His servants had liveries with his crest (a boare) . . . None of his servants durst appear before him without Spanish leather bootes.'

'The world was made for man,' as Bacon himself remarked, 'not man for the world.' The flatterer flourished, as extravagant in the giving as he was in the taking. Yet beneath all the grovelling, and consequent display, there existed a more covert individual. As his physician William Harvey noted: 'He had a delicate lively hazel eie . . . it was like the eie of a viper.' His mother, predictably, had other things to say – reproaching him for not attending church. But she was also aware of a more wayward goings-on at Gorhambury. She remonstrated violently with him over a certain 'proud profane and costly fellow' whom Bacon kept 'as a coach companion and bed companion'. In her Puritan eyes, Gorhambury was filled with 'cormorant seducers and instruments of Satan'. There is no doubt that Bacon was homosexual. The marriage he contracted at the age of forty-five to Alice, a rich alderman's hatchet-faced daughter, was for transparently mercenary purposes. The marriage was never consummated, and Alice was driven to a life of constant infidelity. (Bacon would later cut her out of his will, leaving Gorhambury to his chief steward. No matter: within ten days of Bacon's death Alice had married the steward, eventually driving him 'deaf and blind with too much Venus', according to the fanciful Aubrey.) In vain Bacon's mother pleaded with him to renounce his 'most abominable and darling sin'. In Elizabethan times homosexuality was considered against both nature and nurture: a heinous offence which could have led to something more cutting than public disgrace. Yet Bacon's homosexuality was never used by any of his enemies, of which he had made many on his devious passage to the top. One can only assume that the gossip of history was a well-kept secret in his own time, no matter how unlikely this may seem.

Many have found it difficult to see how such a man could have become a great ornament of a great age – which he undoubtedly was. The life and the inner man (the possessor of the 'viper's eie') appear to have coexisted without changing one another. They certainly impinged on one another (on occasion catastrophically so), but the essential man and the essential mind somehow remained separate. This, as well as the nature of the age, makes it difficult to pass judgement on Bacon's behaviour, even at its most unseemly. No one in the land had a similar hand of talents, and in a different age he might have become lord chancellor on merit. As things were, it is difficult to see how else he could have risen to this top post. And having at last 'come to dignities' he implemented sweeping reforms which had the effect of speeding up the outmoded and corrupt legal system, as well as finding time to dedicate himself to his writings. As we shall see, it was during this period in high office that he produced much of the philosophy for which he is still remembered. Here was the Renaissance man exercising his powers to the full. Yet being a man of the English Renaissance, he preferred to retain a few quaint medieval customs – such as accepting bribes from those whose cases he was trying. As lord chancellor Bacon was the senior judge in the land. This post also meant that he was the senior civil servant. When the king travelled north to visit his native Scotland, Bacon was left as regent. During this time he was virtually king of England.

How are the mighty fallen. In 1621 Bacon was accused of taking bribes. His defence was characteristic of his uncorrupted inner dignity, as well as his deviousness and lack of principle. He freely confessed that he had accepted bribes – going so far as to admit that on occasion he had even accepted bribes from both sides in a case. Yet he insisted that he had never let any of this affect his legal judgement, which had always remained aloof from such mercenary considerations. The authorities and the king were not impressed. Bacon was stripped of his office and flung into the Tower of London. At the same time he was forbidden to hold any further public office, and fined £40,000 (a colossal sum, sufficient to buy four sizeable country estates). But two days later the king had him released from the Tower and remitted his fine. This suggests that he recognized Bacon's sterling qualities and his contribution to his office, realizing that he had in fact been brought down by his enemies.

Bacon retired in disgrace to his estate at Gorhambury, where he devoted his talents to intellectual pursuits. He was now sixty years old, and had five years remaining to him. The works he produced during this period indicate that all his efforts in politics, and even his eventual success, had been little more than a diversion of his supreme talents – the weakness (and waste) of one of the supreme intellects of his era.

Even before his fall, Bacon had produced essays, poetry, philosophy and history of the highest calibre. Indeed, his major work of scientific philosophy, the *Novum Organum*, was written while he was lord chancellor. The title refers directly to Aristotle's *Organon*, the work in which he outlined how knowledge was arrived at by logical deduction. Bacon's aim was no less than to establish an entirely new method of arriving at knowledge. It would supersede the Aristotelian method, which had held good for two millennia, and for the first time establish a firm foundation for the advance of scientific knowledge.

Previously science had been characterized by two approaches. 'Those who have handled science have been either men of experiment or men of dogmas. The men of experiment are like the ant; they only collect and use; the reasoners resemble spiders, who make cobwebs out of their own substance. But the bee takes the middle course; it gathers its material from the flowers of the garden and of the field, but transforms and digests it by a power of its own.' To elaborate: the first method was followed by the 'empirics', who simply built up a jumbled body of unrelated facts. (For Bacon alchemy fell into this category.) The second, the Aristotelian approach, was more systematic but equally misguided. The Aristotelians relied upon deductive logic, where a conclusion follows necessarily from certain premises. For instance, given the two statements:

All planets orbit the sun.
The earth is a planet.

By deductive logic, it necessarily follows that:

The earth orbits the sun.

Here the reasoning moves from general statements to particulars. Bacon was convinced that scientific knowledge could move only in the opposite direction – from particular instances to general principles. Particular instances are tested in experiments, and from these a general theory can be formed. For instance: when observed in a vacuum, objects of different weight always fall at precisely the same rate. From this we deduce that all objects in a vacuum fall at the same rate. Previously Aristotle had maintained that heavier objects fell faster than lighter ones – a plausible conjecture which was accepted for over two thousand years. Not until Galileo conducted his celebrated experiment of dropping objects of different weight from the Tower of Pisa was it disproved. (Though the full proof of Galileo's conjecture would not be possible until such experiments could be conducted in a vacuum.)

Bacon maintained that science could build up a body of knowledge only by inductive logic. This method he characterized as the inferring of general principles from the observation of many particular instances. For example, after observing the sun rise each morning, we induce the principle that it will rise every morning. But even here it was necessary to proceed with caution.

Aristotle had pointed out fallacies in deductive logic. Bacon showed that inductive logic could likewise fall prey to 'false notions' and 'prejudices'. These 'idols of the mind', as he called them, came in four distinct categories.

'The Idols of the Tribe have their foundation in human nature itself . . . human understanding is like a false mirror, which, receiving rays irregularly, distorts and discolours the nature of things by mingling its own nature with it.' For instance, there is a universal propensity for oversimplification. We assume a greater order in things than actually is the case. Likewise, spectacular or sensational occurrences, which may well be unrepresentative, tend to influence our judgement more than routine ones. (Contrary to popular myth, venomous snakes are for the most part retiring creatures.)

'The Idols of the Cave are the idols of the individual.' These are the prejudices and intellectual peculiarities which result from our particular upbringing, education and experience. For instance, when assessing things one person may concentrate on likenesses, another on differences; one on details, another on the whole. Each of us 'has a cave or den of his own, which refracts and discolours the light of nature'.

Idols of the Market Place result from our interaction with others, where 'the ill and unfit choice of words wonderfully obstructs the understanding'. These are the errors due to our use of language. Such errors do not necessarily result from the misuse of language, they may even result from the language itself. 'Words plainly force and overrule the understanding and throw all into confusion, and lead men away into numberless empty controversies and idle fancies.' This understanding of the way language itself can lead us astray was three centuries ahead of its time. Not until Wittgenstein was this problem fully addressed by philosophy. (A single contemporary example will suffice. In the very same year that Bacon was penning these words, Descartes was sitting in his stove doubting the entire fabric of the world and his existence in it. The one thing he found he could not doubt was 'I think therefore I am.' In fact, the process of thinking itself was the one thing he could not doubt – the attachment of the 'I' to this process was imposed on him by the syntax of the language in which he was thinking.)

Bacon's fourth false notion he named 'Idols of the Theatre'. These consisted of 'the various dogmas of philosophies'. He declared, 'All the received systems are but so many stage-plays, representing worlds of their own creation.' He included amongst these idols 'many principles and axioms in science which by tradition, credulity, and negligence have come to be received'. The attack on the unquestioned axioms of Aristotelian science is evident. 'Received' knowledge is antithetical to science, which progresses through discovery. This was an important insight for any new philosophy of science, and it is no coincidence that it occurred in an age when other branches of knowledge were expanding as never before. By now the combination of the Renaissance and the spread of printing presses had spawned a new educated urban population (in London, the audience which appreciated the wit and references of Shakespeare). At the same time physical horizons were expanding too, as Europe embarked on an era of unprecedented geographical discovery. (In 1522 Magellan's expedition completed the first circumnavigation of the globe. By the time Bacon was writing the *Novum Organum* almost a century later, 80 per cent of the navigable coasts of the Americas, Africa and India had been mapped by European explorers, Australia had been sighted and the only major land mass on the globe yet to be discovered was Antarctica.) For the frontiers of scientific learning such miraculous expansion still lay in the future – but Bacon was constructing the navigational aids. Here was the philosophy which individual scientists all over Europe, in their own piecemeal practical fashion, were tentatively reaching towards. Scientific knowledge was cumulative; it was by nature progressive, not conservative.

Yet Bacon well understood that deductive reason could work only if properly applied. It was no good making premature generalizations from just a few cases. Each generalization had to be well grounded in relevant observation. Only then could it be accepted, allowing science to progress by 'gradual ascent' to generalizations of an increasingly general nature. Yet induction always relied upon 'the greater force of the negative instance'. One particular false instance always disproved any generalization. In science, the particular was stronger than the general. This was crucial.

Bacon's inductive method was all very well – but in practice the experimenter often first forms a theory, which he then tests by experiment. The hunch, the intuition, the flash of insight: in reality, this is frequently the starting point. For many years, commentators wondered how Bacon could have missed this simple detail. Only recently has it become clear that Bacon, despite all his insistence, was no great experimentalist himself. He did carry out experiments, but it is now known that many of the experiments he described were 'derived' from other sources – though his accomplished literary flourishes (and highly perceptive insights) often disguised this fact. Yet Bacon's understanding of 'experimentalism' was unparalleled. Take, for instance, his remarks on alchemy. Bacon recognized that the investigation of matter (or what we would call chemistry) was a fundamentally practical pursuit, and that in his time the only people who really knew anything about this field were the alchemists. Thus, if alchemy was ever to be pursued in a rational fashion, and transformed into a science, this would have to be based on the techniques and discoveries of the alchemists. It could not be based upon some imposed rational scheme of principles and laws imported from some other sphere. Here was the way forward for chemistry.

Despite Bacon's profound understanding of experimentalism,

he remained surprisingly out of touch with the experimental advances being made around him. There were now several experimenters of the highest order working in England. Amongst these was the physician and physicist William Gilbert. Following Galileo's Tower of Pisa demonstration, Gilbert again showed how experiment could overthrow authority. According to generally accepted wisdom, garlic had the power to destroy magnetism. Gilbert demonstrated the fallacy of this by simply rubbing a magnet with crushed garlic cloves and showing how it could still attract nails. Such demonstrations were constantly necessary if the progressive theory of knowledge was to overcome the conservative one. It is all but impossible for us to conceive of the profundity of change involved here. A paradigm, a mindset, an episteme – call it what you will – had to be turned inside out. Previously science had been viewed much like a game, such as patience, where each move was made according to a set of rules. But now the rules no longer applied. Imagine how difficult it would be to accept a game where the rules were made up as you went along.

The ancient Greeks had discovered that amber rubbed with fleece is capable of attracting light objects and materials. Gilbert was the first to investigate this property in a scientific fashion, discovering that it was also possessed by rock crystal and certain gemstones. Gilbert named this force 'electronics', after the Greek word *elektron*, which means amber. This is the origin of the word electricity. Gilbert also discovered that compass needles dip vertically, as well as swivelling horizontally, leading him to suggest that the earth itself was a huge spherical magnet, with the compass needles pointing to the planet's magnetic pole.

For many years Gilbert lived alone in his London house, which he used as his laboratory. Half a century after his death this house was burnt down in the Great Fire of London, and many of his notes disappeared. Fortunately, sufficient evidence survived to give an indication of the unsung Gilbert's stature. Two centuries before Faraday, Gilbert had an inkling of the great hidden role electromagnetic forces played in the world. Although Copernicus had convincingly demonstrated that the planets orbited the sun, and Kepler had now shown mathematically that these orbits were ellipses, no one really knew how the planets were held in these orbits. It was Gilbert who suggested that some kind of magnetic force might be responsible. With hindsight, we can see that Newton's momentous idea of gravity, which he produced half a century later, was in a certain way an extension of this simple unverified suggestion floated by Gilbert. His name was to be immortalized in the unit of magnetomotive force, which for many years was known as a gilbert – though today this has been superseded by the amp.

Surprisingly, Bacon criticized and ignored Gilbert's experimental work in equal measure. There are, however, certain understandable reasons for this. It seems that Gilbert's theories went well beyond the results of his experiments, building a scientific–metaphysical construct out of his electrics. Bacon may have believed that the 'electrics' itself was all but metaphysical – a mere fringe phenomenon, of little import to science. It is arguable that Bacon was right to oppose this form of science, even if he was in certain ways mistaken about its contents. Science needed to shed the metaphysical influence of Aristotle, and indeed all metaphysics, before it could proceed. Only then could it arrive at an all-embracing explanation of the world – such as gravity – which resembled metaphysics in its universality.

Bacon's other lacuna is simply staggering. Sir William Harvey, who discovered the circulation of the blood, was for a while Bacon's personal physician. Yet Bacon simply didn't know about Harvey's discovery – which was the death knell of Galen and medieval medicine. It has been argued that Harvey didn't publish his *Exercitatio Anatomica de Motu Cordis et Sanguinis* ('On the Motions of the Heart and Blood') until 1628, two years after

Bacon's death. However, he had been working on this subject for many years, and even began giving public lectures on it a dozen years before he published. One wonders what on earth these two great scientific figures talked about when they met for a consultation. Harvey made his discovery by precisely the techniques of observation and experiment which Bacon had recommended. Galen had taught that the blood oozed back and forth. When Harvey tied an artery he noticed it bulged with blood on the side leading to the heart. When he tied a vein, it bulged on the side away from the heart. He concluded that the blood passed to the heart by way of the veins, was pumped through the heart (rather than seeped through its walls by tiny invisible holes, as Galen had maintained) and flowed away from the heart by way of the arteries.

But the omission was not all on Bacon's side. Harvey esteemed Bacon for his wit and style, but perplexingly concluded: 'He writes philosophy like a lord chancellor.' In other words, his scientific theories (or natural philosophy) were so much pompous nonsense. But Harvey was an odd bird. Years later, during the Civil War, he attended the battle of Edgehill as physician to Charles I. While the battle raged around him, Harvey is said to have passed the time quietly reading a book, waiting for any royal summons.

Bacon, on the other hand, was much more impetuous. When he did undertake his own experiments, he seems to have been something of a bungler. And it was this which led to his demise. In March 1626, while travelling through the snow in his carriage, he had an idea for an experiment. Could refrigeration stop flesh from putrefying? He leapt from his carriage, bought a chicken from a woman at her cottage door and then began stuffing it with snow. As a result he caught a chill, which rapidly became pneumonia. Within a fortnight he was dead.

*

Where Copernicus started the scientific revolution, Bacon set out the mental revolution which would have to accompany it. His style and breadth of mind were to prove an inspiration to the coming generations, which provided the major accomplishments of this revolution. 'If a man will begin with certainties, he shall end in doubts; but if he will be content to begin with doubts, he shall end in certainties.' 'They are ill discoverers that think there is no land, where they can see nothing but sea.' 'Silence is the virtue of fools.' Yet for all his brilliance, it was Bacon's example which was most influential. While he was lord chancellor he was created a peer, and took the title Lord Verulam. If this newfangled science, with its menial experiments, was good enough for a lord, it was good enough for any gentleman. Science became acceptable, even fashionable, amongst the educated English classes. (Yes, once upon a time snobbery actually encouraged science!)

Bacon felt sure that one day science would bring immense benefits to mankind. The early stages of the scientific revolution, during the seventeenth century, produced a number of important inventions – such as the telescope, the microscope and the calculating machine, to name but a few. But for the most part these aided science only; they brought about great achievements in the field of knowledge, but not in the world at large. Science had little effect in these early years. Bacon's idea that science would improve the world was far ahead of its time. (It would be almost two centuries before the steam engine helped bring about the Industrial Revolution.) And like his thirteenth-century namesake Roger Bacon, Francis Bacon also had an uncannily accurate idea of what these benefits would be. In *The New Atlantis*, which was first published after his death, Bacon sketches in some detail his prototype vision of a scientific utopia. The very word utopia had entered the language only half a century previously, when Sir Thomas More published *Utopia*, whose title came from the Greek for 'no place'. More's Utopia had been a social, legal and political paradise. Bacon was the first to realize the role scientific invention would play in such a brave new world.

In *The New Atlantis* the narrator describes how his ship is driven off course and is eventually wrecked on an unknown shore. Here he discovers the New Atlantis, where science has produced spectacular benefits for its inhabitants. There are machines that can travel under water, others that can fly. Medicines have been found that can cure diseases and prolong life. There is artificial lighting; and people can speak to each other over long distances by means of sound carried through pipes. Artificial weather can be produced; natural disasters such as earthquakes and floods can be predicted; animals are crossed to form new species, which are used to test new medicines and chemicals; and buildings have been erected that reach high into the atmosphere.

Yet most intriguing is the element of the vision which he got wrong. Namely, the scientific society. Here, amidst the benefits of science, people lived in harmony. No one stole from his neighbour, or used violence against him. Sex took place only within marriage, and the society was 'free from all pollution or foulness' (alas, he was speaking here in the moral, rather than the ecological sense). Indeed, there was no crime of any sort, and no promiscuity whatsoever.

A likely story. The tenor of Bacon's moral observations may well have been due to his unfortunate prescription for women, which was prophetic only of a fundamentalist Islamic republic. At public events, and even family celebrations, a woman was expected to remain hidden behind a screen, 'where she sitteth, but is not seen'. One can only assume that Bacon's homosexuality and his difficulties with his mother had something to do with this. Yet such attitudes were not limited to men with domineering mothers and a penchant for menservants in 'Spanish leather bootes'. As the historian of science Margaret Wertheim has recently made clear, this baleful and unnecessary strain of misogyny would become increasingly pervasive in scientific circles.

It is difficult to see how someone with such a deep understanding of human deviousness (from inside, as well as out) could have believed all this utopian tosh. But perhaps it was just an idea – the morally beneficial direction in which science was leading us. After all, this was the first comprehensive vision of the scientific future. And it would not be the last to exhibit such naive optimism. Only in the latter part of the twentieth century did we shed the last remnants of our belief in science as a moral force for the good. And even today, when society depends more than ever upon science, we still find it difficult to accept that science itself is morally neutral. It is only human action which invests science with the power for good or evil – utilizing it to create a cure for AIDS or for cloning Saddam Husseins.

However, in its central notion *The New Atlantis* rises to utter fantasy. This society is kept going by a quasi-monastic order of altruistic and utterly unbelievable scientists, who are treated with reverence and respect by the lesser mortals who benefit from their genius. These scientists live and work in peace and harmony in Solomon House, where they are aided by unusually helpful lab assistants, as they all strive to discover new scientific benefits to pass on to society. Ironically, it was this saccharine heart of Bacon's impossible utopian ideal which had the most profound effect. Solomon House was to be the inspiration for the founders of the Royal Society, which, during the following century under the presidency of Newton, would become the leading scientific institution in Europe.

7

A Born-again Science

Science was clarifying its self-image, and even in the the murky realms of chemistry the influence of this process was soon being felt.

The Flemish physician Jan Baptista van Helmont was born in 1577. After taking a degree in medicine and travelling throughout Europe, he retired to his estate at Vilvoorde, just north of Brussels. Here he lived a largely solitary life pursuing his scientific interests, at the same time remaining a devout mystic, believing that all knowledge was the gift of God. He referred to himself as a 'philosopher by fire'.

Although guided by mystical ideas, for the most part van Helmont did not allow them to interfere with his scientific approach to his experiments. This separation of religion and scientific investigation was becoming increasingly prevalent. It is noticeable in Galileo, Descartes and Bacon, who all wrote about science and philosophy taking place in a mechanical world where there was essentially no need for God. Despite this, all three remained believers. As Bacon put it: 'The knowledge of man is as the waters, some descending from above, and some springing from beneath; the one informed by the light of nature, the other inspired by divine revelation.' Their science often clashed with official religion, yet they remained convinced that it was science which was right. Religion would eventually learn to accommodate their findings, they felt. Their attitude was similar to that of the early Muslim philosophers: the laws of nature and mathematics were the way God's mind worked. To understand more about them was to understand more about God.

The deeply mystic van Helmont persisted in this tradition, though he was prone to the occasional lapse. This was not surprising, for his main inspiration was Paracelsus. Van Helmont's penchant for the mystical side of his mentor's activities led him to claim that he had come across 'the Stone', and had even succeeded in using it for transmutation. It is difficult to reconcile such fibbing with his genuine scientific activities, which were a model of exactitude and scrupulous conclusions. Perhaps he felt it added to his mystical status.

Van Helmont's most memorable experiment was a variant on the one undertaken two centuries previously by Nicholas of Cusa. Van Helmont placed 200 lb of dried earth in a large earthenware vessel; this was then watered and planted with a willow shoot weighing precisely 5 lb. The vessel was covered to protect it from accumulations of dust, and daily watering continued, using only distilled water. This emphasis upon exact measurement, clean working conditions and purity of ingredients was certainly influenced by the alchemical practice of Paracelsus (and was thus unwittingly in keeping with Bacon's recommendation that chemistry could advance only by learning from alchemical practice). After five years of watering, the willow had grown into a considerable tree, which van Helmont then dug up and weighed. He found that the tree weighed 169 lb 30 oz. He then dried out the remaining earth, weighed it, and found that it had lost just 2 oz. The mechanism of biological growth was unknown at the time. So van Helmont came to the understandable conclusion that the tree and all its considerable foliage consisted entirely of water. This had simply been converted by the tree into its own substance. As a result, van Helmont abandoned Paracelsus' idea of three elements (mercury, sulphur and salt), rejected the Aristotelian four-element theory from which it had been derived, and went back to the very origins of elemental theory. Van Helmont believed that his experiment with the willow proved that ultimately everything was made of water, the conclusion Thales had reached over two millennia earlier. Van Helmont's experiment is widely regarded as the first application of measurement to an experiment involving both chemistry and biology, marking the inception of biochemistry.

By one of those quirks of human understanding, van Helmont's mistaken conclusion from this experiment led to one of the most significant developments in chemistry. Van Helmont soon decided that although water was the only element, a significant role in the transformation of water must be played by air. This prompted him to undertake an investigation of air and its properties. What exactly was air? Prior to this time no one had approached this question in a thoroughly scientific manner. How could such an insubstantial substance possibly be investigated? Air was air, and that was that. Alchemists leaning over their malodorous bubbling cauldrons had become aware of other 'airs', and had also noticed that certain substances such as perfumes and various oils produced 'vapours'. (Any medieval citizen living in a street with an open sewer would have been all too aware of this: the forward march of science and that of common sense have seldom been in step.) The alchemists recognized that these vapours were not the same as air, and often referred to them as 'spirits'. This name, with its evident metaphysical overtones, soon became attached to liquids that vaporized easily. Habitual usage gradually narrowed this to one of the most volatile liquids in regular laboratory use, namely alcohol. This is the origin of our use of the word 'spirits' to refer to distilled alcoholic drinks.

Van Helmont understood the significance of this somewhat muddled alchemical knowledge concerning 'airs', 'vapours' and 'spirits'. There were a number of distinct air-like substances. In the course of his investigations he conducted an experiment which involved burning 62 lb of charcoal. This gave off vapour which had an identical physical appearance to air, but had very different properties. For instance, when it was collected in a jar a candle would not burn in it. After burning the charcoal, van Helmont was left with just 1 lb of ash. He concluded that the original charcoal had contained 61 lb of this air-like substance, which he named 'spiritus sylvester' (the spirit of wood) – now known to us as carbon dioxide.

Van Helmont found that spiritus sylvester had the same properties as the air-like substance produced by fermentation in wine and beer, the burning of alcohol and several other processes. He concluded that all of these vapours were one and the same. Further experiments led him to realize that there was a wide variety of these air-like substances, each with differing properties. Some were combustible, others had characteristic pungent odours, some were absorbed by liquids. But van Helmont was hindered from carrying out any deep investigation into this apparent variety of air-like substances, mainly because of the lack of any suitably sophisticated airtight apparatus in which to collect and study them. He remained uncertain as to their identity, eventually reaching another mistaken conclusion which was to prove extremely fruitful for chemistry.

Van Helmont decided that these vapours, which remained devoid of shape, colour and sometimes even smell, were in fact a form of pre-matter. They were the formless substance out of which matter was made. According to ancient Greek mythology, the cosmos ('order') had originally been created out of a similar unshaped, unordered substance called chaos – so van Helmont decided to call these vapours or air-like substances 'chaos'. In Flemish the first consonant of this word is pronounced in a heavy guttural manner, and this is the origin of the word 'gas'.

Distinct fundamental principles regarding the nature of matter were now beginning to emerge. First there had been liquids and solids, now there were gases. Van Helmont's experiments led him to the brink of another significant division of matter. In line with his insistence upon experimental exactitude, van Helmont became very keen on the use of balance scales for measuring loss or increase of weight. He discovered that certain metals could be dissolved in various strong waters (acids). These then produced 'savoury waters' (solutions which had colour or taste that could be 'savoured'). Finally, it was even possible to recover from this solution exactly the same weight of metal as had been dissolved into it. This led van Helmont to an understanding of the fundamental property of matter. Despite its transformation during experiments, matter was never destroyed.

Van Helmont's pioneering biochemical work also led him to investigate human digestion. He concluded that 'hungry acid' in the stomach reacted with the food, and that digestion was a process of fermentation (producing the same gas as alcoholic fermentation and the burning of wood). Van Helmont was on the brink of a vital discovery here, but this would only be made some years later by his pupil Franciscus Sylvius, who in 1658 became professor of medicine at Leyden, by then one of the leading universities in Europe.

During the seventeenth century the newly independent Netherlands became a haven of religious toleration and freedom of thought. The driving social force was the rising Protestant middle class, who were more interested in commerce than in waging destructive wars against the Catholics. Here again, as with ancient Greece and the city states of Renaissance Italy, there seems to have been a link between democracy (of sorts) and intellectual development. It was in Holland (and to a certain extent England) that the intellectual renaissance produced its first great achievements. The philosophers Descartes and Spinoza lived there, the English philosophers Hobbes and Locke published there, and the Encyclopedist Bayle took refuge there from the oppressive France of Louis XIV. The local Dutch also contributed to the international intellectual revolution flowering within the safe haven of their borders. The Dutch scientist Huygens would emerge as Newton's only near rival in the field of optics. And Sylvius brought to fruition the wealth of chemical ideas instigated by van Helmont.

Sylvius saw digestion as a 'natural' chemical process, involving acid saliva, alkaline bile and the recently discovered pancreatic juices, which by reaction and taste were deemed to be acid. (As with 'savoury waters', taste remained an important laboratory aid.) Digestion was seen as a chemical warfare. The acids and alkalis produced by the body broke down the ingested food, reacting with the alkalis and acids in it, with these opposing forces eventually neutralizing one another. In the process gaseous effervescence was produced (as we all know), and the heat generated by these reactions warmed the blood. Sylvius recognized that a similar 'fermentation' took place when acidic vinegar was poured on alkaline chalk: gas was produced and the two eventually neutralized one another. By watching these reactions closely, and drawing conclusions, Sylvius extended the boundaries of chemical understanding. He recognized that many naturally occurring salts resulted from the reaction between acids and alkalis. Being composite, these were different from chemicals which could not be broken down. Once again, experimentalism was groping blindly towards an essential distinction (the distinction between elements and compounds as we understand them today).

Sylvius' work was then carried a step further by his pupil, the obscure German apothecary Tachenius – who may well have lived a highly interesting life, before his death in Venice around 1670. (All that is known about him are various details recorded by his enemies, and these are uniformly boring.) Tachenius became convinced that his master Sylvius had overlooked a vital point in his work on acids and alkalis. Here lay the key to it all! Tachenius felt certain that acid and alkali were the two principles which subsumed all chemical reactions. Indeed, he was inclined to believe that these were the two basic elements. This was an important distinction, even if it wasn't quite so fundamental as Tachenius suggested. Unfortunately, at the time it proved difficult to use as an analytical tool, mainly because there was no real definition of what an acid or an alkali was. An acid was only classified in terms of whether it effervesced with an alkali, and vice versa, a purely circular definition.

Despite his ambitious German pupil's attempt to hijack his findings, Sylvius continued to lecture at Leyden, producing experimental and theoretical work of a high order. However, it was his fate to become immortalized for a brilliant stroke of quackery, when he claimed to have invented a cure-all for kidney complaints. This consisted of distilled grain spirit flavoured with juniper berries – which are called *genever* in Dutch. Our abbreviated version of this is 'gin'.

The acid–alkali distinction held the key to a science of practical chemistry. But Sylvius also recognized a further distinction, which would have ramifications for theoretical chemistry, and indeed all natural philosophy. Descartes had viewed the body as essentially a mechanical device. It now became clear to Sylvius that it could equally well be viewed as a chemical device. As a result of the work by van Helmont, Sylvius and Tachenius, chemistry was beginning to emerge as a distinct science.

Even before van Helmont's death scientists had begun to investigate air. As early as 1643 the Italian physicist Evangelista Torricelli conducted a crucial experiment. Put simply, he took a long glass test tube and filled it with mercury. After placing his thumb over the open end, he inverted the tube and placed it in a vessel also filled with mercury. He found that the mercury in the tube sank to a certain level. This was always approximately 76 cm higher than the surface of the mercury outside the tube, which remained open to the air. Torricelli concluded that it must be the weight of the outside air that forced the mercury inside the tube to a higher level than the mercury outside. Torricelli had discovered air pressure. Over the next few days, Torricelli noticed that the height of the mercury inside the tube would occasionally vary by a small amount. He had invented the world's first barometer, for measuring air pressure. News of this soon reached the celebrated French mathematician, religious zealot and hypochondriac Blaise Pascal, who immediately grasped the significance of Torricelli's discovery. He suggested conducting a similar experiment to Torricelli's on the top of the Puy de Dôme, almost 1,500 m above sea level in the mountains of central France. Unfortunately Pascal found that he was unable to conduct this experiment himself, for fear that the mountain air would exacerbate his carefully tended medical condition. So it was carried out by his long-suffering brother-in-law Périer, who discovered that at this height the column of mercury in the tube was considerably less than at sea level – proving that air pressure diminished at higher altitudes. In other words, the amount of air pressing down was less the higher one went. Thus, three centuries before the first rockets were fired into space, it was known that the earth's atmosphere extended only a certain finite distance from its surface. Arguably, Torricelli's experiment had led to the discovery of space itself.

It was now clear that gases were exactly the same as solids and liquids. Like them, gases had weight. They were just much more diffuse (or less condensed). Gases too were a form of matter.

The newly discovered air pressure was to be demonstrated in spectacular fashion by the German engineer and inventor Otto von Guericke, who settled in Magdeburg with his young wife and family in 1627. Four years later he was forced to flee, as the Protestant city was overrun and razed to the ground amidst scenes of savagery and mayhem by the Catholic army of the Holy Roman Emperor. At the end of the Thirty Years War Guericke returned and employed his engineering skills to supervise the rebuilding of Magdeburg, later becoming mayor of the city for over a quarter of a century. Guericke was the first of a new breed: the scientist showman. (A remnant of this breed taught me in the sixth form: he was famous throughout the school for his spectacular hydrogen–oxygen explosion demonstrations, which regularly set off all the alarms and brought in the fire brigade.)

Guericke had the profound intuition and ingenuity of a highly skilled experimenter, and a flair for charismatic presentation which would be the envy of any circus master. But Guericke's 'performances' illustrated serious scientific points. Perhaps understandably in a Germany that had been reduced to a waste-land, the big philosophic topic of the day was the nature of a vacuum. Could such a thing exist? According to Aristotle, it could not, and this had been accepted as 'authority' by later philosophers, who gave us the saying, 'Nature abhors a vacuum.' Guericke decided to settle this question by experiment. In 1650 he invented an air pump consisting of a piston and a cylinder with one-way flap valves. This drew air out of a container in the opposite way to which a modern bicyle pump forces air into a tyre. The power to operate this machine was provided by the local blacksmith (aided later by his assistants, as the going got tougher). Guericke used his new pump to withdraw the air from an iron vessel. He then added insult to injury by resorting to Aristotelian argument to prove that the vessel contained a vacuum. According to Aristotle, if there was such a thing as a vacuum no sound would ever be able to travel through it. Guericke demonstrated that a ringing bell inside his vessel could not be heard. Later, he also demonstrated that a candle could not burn in a vacuum, and that a dog died when confined within one. (Though it would be some years before anyone really understood quite why Rover had become a martyr to science.)

Guericke's most famous performance was carried out on 8 May 1654 before the Emperor Ferdinand III, and attracted vast crowds from all over Saxony. This time his experiment involved two large hollow copper hemispheres, which had been precisely cast so that their rims fitted tightly into one another. (These were to become known as Magdeburg hemispheres.) The emperor sat enthroned on his platform, above the crowd gathered in the sunlit square before the parliament building. All watched expectantly as Guericke greased the rims of the two hemispheres and carefully fitted them together. Then the blacksmith began vigorously pumping the air from inside the sealed copper globe. After a while he was joined by his assistants, as the cranking of the pump gradually became more and more laborious. The crowd then watched in bemusement as a team of eight horses harnessed together was led into the square and attached to one hemisphere of the copper globe. Another team of eight horses was then attached to the other hemisphere. At a signal from Guericke, the two teams of horses strained forward in opposite directions, attempting to pull the two hemispheres apart. The crowd fell silent as the powerful drayhorses heaved, but no matter how they were whipped they were unable to separate the hemispheres. Guericke now addressed the emperor and the crowd. This was not a trick, he told them. All that was holding together the two hemispheres was the pressure of the air surrounding them. The vacuum inside the globe meant there was no opposing pressure to balance this great outer force – which was even more powerful than the strength of sixteen horses. The emperor was amazed, an emotion that was clearly echoed on the gaping faces of his gathered subjects, who now began to applaud. But Guericke held up his hand, silencing the crowd. The experiment was not over yet. The horses were untied from their harnesses and led away. Guericke then fiddled with the pump. The crowd craned forward curiously. There was a sudden hissing sound as the outside air pressure rushed into the hollow globe to fill the vacuum. Then, without warning, the two copper hemispheres simply fell apart of their own accord. Now that there was no vacuum, and the pressure inside was the same as the pressure outside, there was nothing to hold them together.

Guericke's experiment soon became so famous that he was demonstrating it all over Germany. Tales of his feat with the Madgeburg hemispheres spread across Europe; etchings were made of it; and when a garbled version crossed the Channel to England, it is said to have given rise to the nursery rhyme about Humpty Dumpty:

> Humpty Dumpty sat on a wall,
> Humpty Dumpty had a great fall.
> All the king's horses and all the king's men
> Couldn't put Humpty together again.

Not for the first time, or the last, the English appear to have completely misunderstood what was going on in Europe.

As a result of experimental work by the likes of van Helmont, Torricelli and Guericke, the unknown properties of matter, to say nothing of its unsuspected hidden powers, were at last beginning to emerge. The public imagination was intrigued, just as the private imagination of many gentlemen with inquiring minds in England, and elsewhere, had been inspired by the natural philosophy of Milord Verulam (Francis Bacon). It looked as if an entire science lay hidden in matter itself. The field of chemistry was now ripe for the intervention of a great scientific mind. This duly appeared in the form of Robert Boyle.

Considered by many as the founder of modern chemistry, Robert Boyle was born in 1627 in a remote castle in the south-west of Ireland. He was the fourteenth child of the irascible, ageing Earl of Cork, who, as lord chancellor of Ireland, was busily acquiring a string of estates – which would eventually stretch unbroken from the Irish Sea to the Atlantic. (The practices which had recently put paid to Bacon's lord chancellorship in England were still flourishing across the water.)

Young Robert quickly gave evidence of an exceptional mind, becoming fluent in Latin and Greek by the age of eight. Whereupon he was dispatched, along with his twelve-year-old brother, to England to study at Eton. Here the child prodigy suffered from a combination of sickliness, a nervous stutter and the educational method favoured by England's most prestigious public school: regular enthusiastic caning for all pupils, regardless of age, social standing or intellectual ability. As a result, the traumatized Boyle 'forgot' most of his Latin and was reduced to bouts of suicidal gloom, which continued until he was a teenager. By this stage Robert and his older brother Francis had been sent off with a tutor for a prolonged tour of the Continent. At the age of fourteen, while in Geneva, Boyle had an experience which was to transform his life. In the middle of the night he was woken by a violent summer thunderstorm, which soon took on eschatological proportions in the young Boyle's mind. Quaking in his bed, he imagined outside the shutters 'Sallys of that Fire that must consume the World . . . Apprehensions of the Day of Judgement's being at hand.' He was seized with the fear that his soul was in no fit state to encounter a wrath even greater than his father's or that of an Etonian schoolmaster. In his anguish he vowed 'that if his fears were that night disappointed, all his further additions to his life should be more religiously and watchfully employed'. Tellingly, Boyle refers to himself in the third person throughout this memoir of his troubled early years. His teenage vow sounds comparatively reasonable, but it was not. These words were accompanied by a surge of fervent religiosity and the usual vow of chastity. Most such youthful impetuousness is soon forgotten, but Boyle's religiosity and chastity were to remain a permanent feature of his life.

This combination of fanatical devotion and original scientific thinking was not uncommon during the period. And such obsessive faith was no self-protective affectation of genius either. Van Helmont, Pascal, Spinoza and Newton all considered that their religious thought was their major contribution. A curious aberration, which caused them to suppose they would not be remembered for their philosophy, mathematics or science (amongst the finest ever produced), but for their theology (which for the most part was certainly not). Likewise, they all appear to have remained celibate. While many may consider this latter non-behaviour pattern a harmless foible, no business of anyone else, its effects give the lie to this attitude. Combined with Bacon's heterophobia, the touchy celibacy, misogyny and sexual hypocrisy which prevailed amongst the scientific community (especially in England) were to have an incalculable effect on the new revolution in knowledge. Women became so feared that they were ruthlessly excluded. Half the population was prevented from making any contribution whatsoever. One has only to think of the gaffes which might have been averted, ground-breaking theories which might have appeared earlier . . . Double the workforce could have been applied to these problems.

Robert Boyle continued on his grand tour, all the while receiving excellent instruction from his private tutor. Boyle was pleased that his tutor paid much more attention to his 'Scholler's Proficiency, than the Gaines he might deriue from the common tedious & dilatory way of teaching' he had experienced at school. Freed from the constraints of syllabus or spelling, Boyle was able to follow his own reading pattern. His precocity returned, this time in scientific guise. In Italy, he was able to read the works of Galileo just a few months after he had died. From this early age Boyle realized the fundamental importance of experiment. He also read Descartes, absorbing his mechanical world-view. Had Boyle been educated at almost any university in Europe, such instruction would not have been available. Even at this late stage, Aristotelianism remained the academic order of the day. Boyle's lack of orthodox formal education was to prove the making of him. But even a celibate young swot must have his diversions. Arriving in Marseilles he missed a boat trip to the coral reefs, which in those dolphin days were the main tourist attraction. However, he 'had the pleasure to see the King's Fleet of Galleys put to Sea, & about 2000 poore slaues tugge at the Oare to row them'.

Boyle returned to England to find the country plunged into conflict between the Parliamentarians and the Royalists. The rising middle class, which sought more democratic power for Parliament, was opposed to the monarchy, which believed in the 'divine right of kings' to rule as it saw fit. The Civil War would result in the execution of Charles I in 1649, and the establishment of the Commonwealth: the first successful revolution in Europe. Boyle's family were for the most part Royalists, though his favourite sister, Katherine, was an ardent Parliamentarian. Boyle himself did his best to avoid the war, settling in rural Dorset.

By now Boyle was a tall, rather frail young man with poor eyes, his long gaunt features framed by a shoulder-length curled wig, in the fashion of the period. In Dorset Boyle embarked upon his first serious chemical experiments, and also wrote essays – one of which is said to have inspired Jonathan Swift to write *Gulliver's Travels*. Occasionally Boyle would travel to Ireland to administer the estates he had inherited from his father. During these trips he was unable to continue with his chemical experiments, for the simple reason that there was no chemical apparatus in Ireland. In 1656, at the age of twenty-nine, Boyle moved to Oxford.

This city had been a Royalist stronghold during the Civil War, attracting many important refugees from Parliamentary London. Amongst these were a number of scholars who were interested in the new experimentalism, which was now fashionable as a result of Bacon's writings. These, and other natural philosophers, took to meeting on an informal, irregular basis to discuss the latest scientific undertakings. Boyle was soon attending the meetings of this group, which served as a focus, encouragement and information exchange. The group became known as the Invisible College. In 1662, two years after the restoration of Charles II, this group of largely Royalist scholars was granted a charter, and became known as the Royal Society. The society nailed its anti-Aristotelian colours to the mast with its motto: '*Nullius in verba*' (nothing by word alone, i.e. nothing by mere authority). It would place its faith in the scientific approach, with experimentalism being the order of the day.

In Oxford Boyle took up residence at a house in the High Street, where he set up a laboratory. To assist him in his experimental work he hired a touchy, impecunious young Oxford undergraduate with a pock-marked face called Robert Hooke. Despite their differences, the twenty-nine-old aristocrat and the difficult twenty-one-year-old clergyman's son immediately formed a profound working relationship, which was to last a lifetime. For the first time, Boyle had met his intellectual equal.

Hooke would go on to become an outstanding physicist, overshadowed only by Newton, a state of affairs that was to cause Hooke great chagrin. Several years before Newton published his epoch-making work on gravity, Hooke wrote to Newton suggesting a gravitation theory of his own – though in the event this proved to be flawed, and had little real mathematical backing. In a different field, Hooke's pioneering work with the microscope on living organisms led him to coin the word 'cell'.

Boyle had heard of Guericke's public experiments with the

Robert Boyle

Magdeburg hemispheres, and with the aid of Hooke he set about designing an even better air pump, which was attached to a vacuum flask. Using the latter, Boyle proved Galileo's prediction that in a vacuum any two bodies would fall at precisely the same rate. In the absence of any air resistance, a feather fell at the same rate as a lump of lead, contrary to Aristotle's assertion. Boyle and Hooke also confirmed Guericke's finding that sound didn't travel through a vacuum. But they did make two surprising discoveries: electrical current could be felt across a vacuum, and insects didn't die in one. However, other animals did. These experiments on the absence of air led Boyle to wonder about the nature of air itself. Conducting experiments with mice and birds, he came to the conclusion that air was inhaled and then exhaled from animals' lungs to remove impurities from the body. Boyle conducted his experiments in a methodical manner, aiming to investigate and exhaust all possibilities. In this way he proved once and for all that air was no mystical entity which somehow pervaded the entire world. Air was not an essential element of nature as the Aristotelians believed; it was simply a substance with definite properties of its own. For instance, it made iron rusty, and turned a copper dome green. Also, when compressed, it seemed to have elasticity. Boyle concluded at this stage that air was like an elastic fluid with reactive particles floating in it.

Van Helmont and others had noticed that there were different gases (such as spiritus sylvester), but Boyle was the first to collect these and study them as entities entirely separate from air. He noticed that they all possessed this property of elasticity. At the time Hooke was conducting his famous experiments with metal springs, which would give rise to Hooke's Law concerning the elasticity of bodies. This prompted Boyle to call this gaseous elasticity 'spring of the air', and he now devised an experiment to measure this effect.

Boyle took a seventeen-foot-long J-shaped glass tube which was sealed at the lower end. Using mercury, he trapped some air in the sealed end. He then found that if he doubled the weight of mercury in the tube, the volume of the trapped air was halved.

If he tripled the pressure, the volume was reduced to a third. But if he halved the pressure, by removing half the mercury, the air doubled in volume. From this he induced Boyle's Law, which states that the volume of a gas varies inversely according to its pressure. Boyle now realized that, because it was possible to compress a gas, it must consist of separate particles moving in a void. When the pressure was increased, this merely squeezed the particles closer together.

Fifteen years later Boyle's Law was also discovered independently by the French scientist-priest Edmé Mariotte, but with the important added proviso that the temperature of the gas must remain constant throughout. If a gas is heated, it expands of its own accord; and it contracts when it is cooled. Boyle almost certainly realized this, but with untypical carelessness neglected to mention it in his report on his experiment. A salutory lesson: for this reason Boyle's Law is known as Mariotte's Law throughout Europe.

In fact, it is now known that both Boyle and Mariotte were preceded by the last of the great ancient Greek scientists, Hero of Alexandria, who lived around the first century AD. Contradicting Aristotelian ideas about the all-pervasive four elements, Hero demonstrated that air was in fact a separate substance. He showed how water was incapable of entering an upturned glass filled with air: the water rose up the glass only as the air was allowed to escape. He noticed too that air was compressible, and from this he deduced that it must be made up of individual particles separated by space, just as Boyle would do one and a half millennia later. Hero also realized that when steam was heated, the particles of which it consisted became more agitated. In other words, when steam was heated its pressure increased. This is the principle of the steam engine. Hero quickly realized the possibilities of this source of power, and even demonstrated it by devising the world's first steam engine. This consisted of a hollow sphere containing some water, with two bent tubes inserted into either side of the sphere. When the water in the sphere was heated to boiling point, steam rushed out of the tubes causing the sphere to rotate. (This principle is still used in rotating water sprinklers for lawns.) Hero's machine was not only way ahead of its time, but, like many such inventions, it was deemed utterly worthless and unnecessary by his contemporaries. What on earth was the point of developing a machine for doing work when you could get a slave to do it far more easily?

Boyle's originality shows in the way he developed these independently rediscovered ideas. If air and other gases consisted of particles separated by a void, what about liquids and solids? When water evaporated, it was becoming a gas, particle by particle. Water vapour behaved like any other gas, which meant that it too consisted of particles separated by a void. If this was the case when water was a gas, then it would seem likely that it was also the case when water was a liquid, and even when it was solid ice. And if this was true of water, then possibly it was the case with all substances. Boyle's reasoning was tenuous, and had little or no experimental backing – but it confirms his qualities as an original scientific thinker. Without being fully aware of what he was doing, Boyle was preparing the ground for the reintroduction of atomic theory.

Boyle's masterpiece was *The Sceptical Chymist*, which was published in 1661. This is generally regarded as the beginning of the new chemistry. Indeed, its very title led to the widespread dropping of the 'al' from alchemy, to give chemistry. The new science would shed its oriental esoteric past.

Around this time Boyle also initiated the practice of writing up his experiments in a clear and easily comprehensible manner, so that they could be understood, repeated and confirmed by other scientists. This was the very opposite of alchemical secrecy, and proved a major advance for science as a whole. By setting this example, Boyle was doing the same for science as the ancient Greeks had done for mathematics. The truths of mathematics, arrived at by deductive reasoning, were confirmed by proof. Now the truths of science, arrived at by inductive reasoning, had their own means of verification. From a secretive practice in a murky den, the new chemistry was transformed into a universal science which could be practised in any laboratory anywhere.

The Sceptical Chymist launches into an attack on the Aristotelian theory of the four elements, and also its Paracelsian derivative of three elements. Instead, Boyle asserts that elements are primary particles. In the words of his famous definition, what he meant by elements were 'certain primitive and simple, or perfectly unmingled bodies; which not being made of any other bodies, or of one another, are the ingredients of which all those called perfectly mixed bodies are immediately compounded, and into which they are ultimately resolved'. In other words, any substance which could not be broken down into a simpler substance was an element. Here for the first time is an understanding of the elements which matches the idea we have today.

Boyle then went on to make a further fundamental distinction. These elements could combine together in groups or clusters to form a compound. (This is the first occurrence of the notion which would develop into the modern idea of a molecule.) Boyle was not just theorizing here. His long experience and expertise in the laboratory had led him to realize that such clusters of molecules existed in the form of stable compounds. (For example, when iron was dissolved in an acid and gave rise to a compound salt, this was a stable substance, yet it could also be broken down and the iron recovered.) Boyle concluded that all these compound substances depended for their properties upon the number and position of the elements they contained. Again, this description is uncannily accurate: the flash of insight which would later lead to molecular theory.

Despite his knack of brilliant theorizing, Boyle, like his colleagues, believed fundamentally in experimentalism. '*Nullius in verba*', as the Royal Society decreed. Not in books, but in the laboratory. They rejected all systems, such as the four elements. Yet Boyle, and his colleagues, did in fact believe in their own overall system. This decreed that the world was composed of corpuscles, which behaved in a mechanistic fashion. Although arguably less metaphysical than Aristotelian notions, this mechanical-corpuscular system was nonetheless fundamentally metaphysical itself. How could such an idea ever be tested in a laboratory? It was, strictly speaking, unscientific. Yet it did have one compelling advantage – which was to be adopted by science from this time onwards. Unlike the four elements, the mechanical-corpuscular system explained a huge variety of scientific phenomena. In other words, it worked (even if science didn't yet know quite why).

However, when examined closely, Boyle's notion of an 'element', together with that of a compound, only hints at the way ahead towards our modern ideas. It wasn't fully articulated. Why not? According to Boyle's definition, an element was a substance which couldn't be broken down into even more primary substances. This meant that when a substance was found to be an element, this could only be a provisional state of affairs. It was always possible that someone would find a way of breaking down the substance still further. (Not until twentieth-century nuclear science would chemistry be provided with a cast-iron definition of precisely what an element is.)

This left Boyle in an anomalous position. Although he had defined an element, he didn't actually know what one was. Ironically, it was just possible, as chemical techniques improved, that all the substances hitherto thought to be indivisible elements would eventually be broken down into just four elements – very similar to earth, air, fire and water! Boyle was convinced that this was not the case, and it certainly appeared unlikely. But this anomalous position led Boyle into a grotesque misunderstanding – one which could well have undermined his entire contribution to science. As a result of his laboratory experience, Boyle formed the conviction that metals were not in fact elements. He had yet to find a way of breaking them down into their primary substances, or elements, but he felt sure that this would one day prove possible. Now this coincided precisely with what the alchemists believed. If metals were not primary substances, one could be broken down into its constituent elements and reassembled – or transmuted – into another. Lead could become silver, silver could become gold.

It would be reassuring if at this point one could say that Boyle realized the theoretical possibility of such alchemical pursuits, and just left it at that. But no. Like Paracelsus, Bruno, van Helmont and so many other chemical pioneers before him, Boyle too caught the alchemy bug. For many years historians of science preferred to skate over this unpalatable fact, mentioning it only in passing – as if it was some unfortunate occupational hazard of chemists at this time. As indeed it was. It is difficult to imagine Galileo, Descartes, Spinoza or Pascal indulging in such practices; physicists, mathematicians and philosophers appear to have remained immune from this particular metaphysical contagion. (Though this would not always be so, as we shall see.)

Sadly, recent close examination of Boyle's papers indicates that his alchemy was no whimsical aberration. He undeniably believed in what he was doing. Secret, coded notebooks reveal a persistent and extensive quest for the philosophers' stone. Though unlike van Helmont, he never actually claimed to have found this elusive entity. On the contrary, Boyle goes about his alchemical experiments with his usual scrupulous scientific rigour, attempting one by one to repeat any allegedly successful alchemical experiments which come to his notice. Appealing to inductive reason, he argues that if he is successful 'this one positive instance will better prove the reality of what they call the philosophers' stone, than all the cheats and fictions, wherewith pretending chemists have deluded the unskilful and the credulous'. Boyle was far too good a scientist to be taken in by all of alchemy: 'for so the writings present us, together with diverse substantial and noble experiments, theories which either like peacock's feathers make a great show, but are neither solid nor useful, or else like apes, if they have some appearance of being rational, are blemished with some absurdity or other that, when they are attentively considered, makes them appear ridiculous'. But his belief in the 'substantial and noble' aspect of alchemy remained. He even used his influence in the Royal Society to press for the repeal of the anti-alchemy law passed four centuries previously by Henry III, which expressly forbade the manufacture of gold by transmutation. Indicatively, Boyle was not against this law because of the futility of the activity it banned. On the contrary, he felt that the manufacture of gold by this method would be of great benefit to the country. Inspired by the customary combination of ignorance and avarice, Parliament duly repealed the anti-alchemy law in 1689. Boyle immediately encouraged scientists throughout the land to pursue this important quest.

The key to Boyle's puzzling obsession appears to lie in his fervent religiosity. As well as natural philosophy, Boyle also published several works on religion. He used his private fortune to finance translations of the Bible into Irish and Turkish – to dissuade the Irish from Catholicism and the Turks from Islam, both equally pernicious heresies in his view. Likewise, he set up a fund to establish the regular Boyle Lectures, which are delivered at the Royal Society to this day. Despite their venue, the declared intention of these lectures was 'for proving the Christian religion against notorious infidels'.

Boyle believed that the world worked according to 'those two grand and most catholic principles, matter and motion'. This clockwork universe had been set in motion by the Creator. Thus the scientific study of nature was a religious duty. But God and the human soul were incorporeal, which set them apart from the mechanical-corpuscular world. However, Boyle seems to have believed that the philosophers' stone somehow interacted with both these worlds. It could attract angels, for instance. And if found, it would prove an effective weapon against atheism. As Boyle grew older he became increasingly obsessed with atheism, evidence of which he was continually detecting in public places and in the trends of society at large. This meant that the pursuit of alchemy was also a religious duty, just like science.

In 1668 Boyle moved to London, where he lived with his favourite sister, Lady Katherine Ranelagh, in a house on Pall Mall. At this time Pall Mall was a leafy suburb between the orchards of Bond Street and the royal gardens. Nell Gwynn, the king's mistress, lived a couple of doors along, from where she would 'draw attention to herself' when Charles went walking with his dogs in the royal park. By now Boyle had achieved great renown. He had even been offered the presidency of the Royal Society, but had been forced to refuse this great honour because the nature of the president's oath conflicted with his religious principles. His house was a mecca for distinguished foreign visitors, such as the German rationalist philosopher Leibniz, and even a Chinese dignitary who had been converted to Christianity. Leibniz, the Leonardo of the intellectual renaissance, revered Boyle but scolded him for not publishing enough of his work. Despite Boyle's diligence in writing up his experimental work, he was curiously haphazard when it came to publication. His influence, and his stature, would have been considerably improved if only he had devoted more of his time to a systematic exposition of his findings.

To the end, Boyle's work remained of the highest order. His greatest contribution to the chemistry laboratory was his discovery of a method for distinguishing between acids and alkalis. He found that syrup of violets turned red with acids, green with alkalis, and remained the same with neutral solutions. At last a proper definition of acids and alkalis had been discovered. Typically, this had been found in the laboratory, rather than worked out in theory. Chemistry's most influential analytic distinction was a working definition.

By the time Boyle died in 1691, at the age of sixty-four, Newton had already published the *Principia*, his revolutionary work on gravity. This produced the first comprehensive explanation of how the mechanical-corpuscular world worked, the forces that held it together. But gravity dealt only with the physical properties of matter. Newton was well aware of this, and sought to remedy the lack. In his work on light, *Opticks*, he appended a number of queries relating to a wider range of scientific topics. These pointed the way forward by suggesting future lines of inquiry for science, including chemistry. 'Have not the small Particles of Bodies certain Powers, Virtues, or Forces, by which they act at a distance . . . upon one another for producing a great Part of the Phaenomena of Nature?' These 'attractive Powers' appeared to be more than the forces by which 'Bodies act upon one another by the Attractions of Gravity, Magnetism and Electricity'.

The trouble was, Newton's ground-breaking scientific vision of the world had set a new standard for science. Any scientific advances were expected to incorporate the mechanical rigour and mathematical exactitude of Newtonian physics. Yet Newton was dealing with the world of quantity, rather than the world of quality to which chemistry still belonged. And even in its fundamental concepts chemistry still remained largely unquantified. Boyle's idea of the chemical element had only just emerged, and was far from being fully understood. Under such circumstances one couldn't expect exact quantitative analysis of the complex chemical reactions these elements underwent.

So much for Newton's orthodox approach to chemistry, whose import still lay in the future. His other approach to the qualitative problems of matter was a complete and utter disaster. It is no exaggeration to say that this was probably the greatest waste of a great mind in the history of science.

Newton plunged into alchemy. Unlike Boyle, whose alchemical pursuits had a smidgeon of theoretical scientific justification, Newton's had none. His headlong descent into the nether world of alchemy was metaphysical from start to finish. From inspiration to desired outcome, the aim was never scientific. Like Boyle, Newton too was obsessed with religion. His particular field of interest was the Holy Trinity, which he believed to be a misconception. This was serious heresy. (Though it also had its less serious side. Newton belonged to Trinity College, Cambridge. As a fellow of Trinity he was required to swear on oath that he would uphold the college and what it stood for. Newton tactfully chose to swear his belief in the existence of his employers.)

At least half of Newton's intellectual life was wasted on non-scientific pursuits. Along with his heretical ideas about the Holy Trinity, Newton spent years on abstruse mathematical calculations concerning the Old Testament and mythology – coming up with precise dates for such events as the creation, Noah's ark and the voyage of the Argonauts. He also learned Hebrew so that he could read through Ezekiel verse by verse in order to reconstruct an exact blueprint of the Temple at Jerusalem. (Though he was willing to switch to Latin or Greek translations when these suited his purposes.) Then, using the symbolism embedded in the Temple, and descriptions in the Book of Revelation, he found he was able to prophesy the exact dates of such events as the Second Coming of Christ and the end of the world. One of the transcendent mathematical minds of all time was convinced that this was the major intellectual work for which he would be remembered by a grateful posterity.

All this was intimately connected to Newton's somewhat lesser alchemical aspirations. Alchemy was where the 'Magick' of the spiritual realm entered the world; to understand this was to understand the workings of the spiritual world. Beside the enormity of his prophecies, Newton's alchemy might seem a mere peccadillo. Until the facts emerge, that is. Newton possessed no fewer than 138 books on alchemy, and was to write over 650,000 words on the subject. This was evidently no mere hobby. His experimental involvement was similarly gargantuan. Despite his secretive approach to the subject, Newton went so far as to construct his own furnace in the garden of his college rooms, in order to proceed with his alchemical investigations. According to his despairing assistant, he was 'so serious upon his studies that he ate very sparingly, he often forgot to eat at all . . . very rarely went to bed, till three or four of the clock, sometimes not till five or six . . . he used to employ about six weeks in his laboratory, the fire scarcely going out either night or day, he sitting up one night, as I did another until he had finished his chemical experiments . . . What his aim might be I was unable to penetrate into.' Judging from Newton's notebooks, his motives for these experiments were wholly metaphysical. Even the transmuting of base metals into gold was regarded from the metaphysical angle. This was not chemistry, it was sorcery.

And yet. Newton was convinced that some kind of structure lay at the heart of matter. There had to be ultimate laws governing this substance, the very stuff of the world we inhabit. But as we have seen, the science of chemistry was in fact not yet sufficiently advanced to tackle such questions – unlike physics, where Newton had succeeded as no other. And of course Newton's chemical motives were unscientific, his mental approach questionable to say the least. Newton's mental stability was precarious at the best of times, but it was the failure of his alchemical investigations, combined with an unadmitted infatuation with a young Swiss mathematician, which finally drove him over the edge. In 1693, at the age of fifty, Newton suffered from 'a distemper that much seized his head, and left him awake for about five nights altogether', according to a Cambridge colleague. He then fled from Cambridge and vanished altogether for several months. The next that was heard of him was a scrawled, ink-blotched letter written at the Bull Tavern in Shoreditch, east London, to his friend the philosopher John Locke. In it he apologizes for 'being of opinion that you endeavoured to embroil me with woemen'. Such an endeavour would have particularly vexed Newton, who lived a life of strictest celibacy, thus ensuring that he didn't have to admit his repressed homosexual inclinations even to himself.

All this had its effect on the scientific world at large. Newton's presidency of the Royal Society ensured that misogyny was enshrined in this venerable scientific institution. This tendency had already been encouraged by Hooke, the secretary of the society, who had taken a vow that he would never marry. It seems that for members of this august institution even the slightest prospect of being embroiled with a woman was intolerable. As the modern historian of science Londa Schiebinger observed: 'For nearly three hundred years the only female presence at the Royal Society was a skeleton preserved in the society's anatomical collection.'

The path of alchemy led to lunacy – not only for chemistry. The elements which Boyle had defined would now have to be discovered in a strictly literal sense.

8

Things Never Seen Before

As we have seen, nine elements were already known to the ancients, and three new ones were discovered in the late Middle Ages. Yet it is of course only with hindsight that we can recognize these as elements. Their discoverers did not see them as such, because they didn't know what an element was. It was only in 1661 that Boyle came up with his definition of an element as a substance which could not be broken down into simpler substances.

Around eight years later the first new element was discovered by Hennig Brand in Hamburg, when he isolated phosphorus. This was a momentous event in the history of chemistry – and not only because it was the first discovery of a new element since the Middle Ages (which Brand couldn't have known). More significantly, it was the first time an element had been discovered which had not previously existed in its isolated state. Here was something which had never before been seen on earth. (Phosphorus does in fact occur free on earth, but only in the occasional meteorite.)

Hennig Brand was something of a rarity himself. He is variously dubbed as the last alchemist or the first chemist. Born in Hamburg early in the seventeenth century, he served as a junior officer in the army, possibly at the end of the Thirty Years War. On leaving the army he set up as a doctor, despite having no qualifications. One source describes him as 'an uncouth physician who knew not a word of Latin'. Fortunately Brand married a rich woman, which enabled him to pursue his real interest – tinkering about in his laboratory. Whether this tinkering involved simply messing about with old alchemical recipes or demanded skilful chemical analysis is still a matter of dispute. Either way, Brand certainly subscribed to the Baconian view that there was a lot to be learned from the alchemists. Like so many before him, Brand even suspected there might possibly be a grain of truth in this transmutation business. He began studying the Paracelsian doctrine of signatures, which suggested that Nature revealed its secrets in symbolic form. According to this way of thinking, a natural object which was gold in colour could well contain gold. By a stroke of serendipity (or ingenious insight), Brand came to associate this idea with a piece of ancient alchemical lore which stated that the philosophers' stone for making gold was contained in the dregs of the human body. Eureka! There was only one possible substance which conformed with both clues – urine.

Brand began a prolonged and extensive investigation of the properties of human urine, which must have sorely tried the patience of his well-to-do wife, to say nothing of the neighbours. He collected fifty buckets of human urine, which he then allowed to evaporate and putrefy until they 'bred worms'. This he then boiled until there was a pasty residue. When he left this in the cellar for some months he found it had fermented and turned black. Doubtless with the entire neighbourhood now up in arms, Brand proceeded to heat the black fermented urine concentrate with double its weight of sand in a retort, whose long neck was plunged into a beaker containing water. The final distillate collected under water in the beaker was a transparent waxy substance. When removed from the water it glowed in the dark, and sometimes even spontaneously ignited, giving off dense white fumes. He decided to name this new substance phosphorus, from the Greek *phos* ('light') and *phoros* ('bringing').

Difficult though it may be to believe, the prolonged experiment described above was one of the best-kept secrets of seventeenth-century science. Brand proudly demonstrated his new substance to his friends in Hamburg, but refused to divulge the secret of how he had produced it. News of Brand's discovery, and the miraculous powers of this new phosphorus, soon began to spread throughout Germany.

Guericke's spectacular demonstrations with the Magdeburg hemispheres had set off a popular craze for scientific experiments in Germany. By now a number of chemists were making a living touring the various courts, demonstrating the latest scientific wonders. Phosphorus was ideal for such performances. Others even felt there might be a military use for this new substance.

Brand was eventually visited in Hamburg by a certain Dr Johann Krafft of Dresden, who persuaded him to part with the secret of phosphorus for 200 thalers. Whereupon Krafft began giving phosphorus demonstrations at courts all over Europe. When he visited the court of Charles II in London, Boyle was invited along to view the experiment. Krafft refused to divulge the secret to Boyle, but by dint of a casual remark Krafft let drop, and careful observation of his experiment, Boyle gleaned certain clues. Within a few months, aided in part by a German chemist called Ambrose Godfrey Hanckwitz, Boyle had independently discovered how to prepare phosphorus. He wrote out a clear step-by-step description of his experiment, sealed it in an envelope and lodged this with the Royal Society. Meanwhile Godfrey Hanckwitz dropped the German part of his name, set up shop in London and began producing phosphorus for distribution all over Europe. Within a few years he had made a fortune and become so famous that a letter simply addressed to 'Mr Godfrey, Famous Chemist in London' would reach him.

In keeping the process a secret, Boyle greatly helped Hanckwitz (who had of course helped him). However, this action of Boyle's was uncharacteristic. His practice of writing up all his experiments in clear and comprehensible fashion was part of his wish that all knowledge in science should be shared, preferably through such institutions as the Royal Society. In this way scientists everywhere could benefit from the latest advances: Bacon's utopian dream, Solomon House, would be realized in scientific institutions all over Europe.

This was all very well for Boyle, who was a rich man with a private income, but others not unnaturally sought to gain from their discoveries. Was science to become commerce, or should it be for the benefit of all? This question remains relevant even today (for example, regarding the patents for the genes of newly 'created' animals). Such difficulties were also compounded by questions of priority. The scientific revolution of the seventeenth century meant that an increasing number of scientific minds found themselves confronted with the same problems, coming up with their own similar solutions and discoveries independently of one another. Who had discovered what first? If a person published his (or her) discovery he (or she) got the credit, but everyone then knew about it. If he didn't publish, then his discovery might be made independently by another, who could then claim the credit.

The laws of patenting remained in their infancy. Previously, the exclusive right to some process would usually be granted by the reigning monarch or ruler. Galileo had been granted exclusive rights by the Doge of Venice to a water pump he had invented. These rights were granted in perpetuity, but only within the Venetian Republic. With the spreading of the scientific revolution throughout Europe, such arrangements became increasingly unfeasible. In 1623 the British government passed a Statute of Monopolies which decreed that 'letters patent' could be granted for a period of up to fourteen years covering 'inventions of new manufactures'. Meanwhile scientific academies had sprung up in Europe. The Royal Society was pre-dated by the Accademia dei Lincei (Academy of the Lynxes) in Italy; later the Académie Royale des Sciences in Paris developed from the informal meetings attended by the likes of Descartes and Pascal; and Leibniz was instrumental in setting up the Berlin Academy. Scientists would demonstrate their new discoveries before the assembled members of these academies, and to a certain extent such institutions acted as repositories and coordinators of new knowledge. But this became increasingly difficult when such knowledge was theoretical, or remained unpublished.

The most notorious instance here was undoubtedly the priority dispute between Newton and Leibniz over who had invented calculus. When Newton published his *Opticks* in 1704, he added an appendix describing the 'method of fluxions' (calculus) he had invented thirty years beforehand. Unfortunately, his arch-rival Leibniz had published his own version of calculus twenty years earlier. Accusations of plagiarism began to fly back and forth. Sadly, the reactions of both claimants soon sank woefully below the level of behaviour expected of genius (which is not always high at the best of times). The facts were simple. Newton had indeed shown Leibniz some earlier papers which alluded to his method of fluxions. But Leibniz' calculus was crucially different, not least in its notation (which is the one we use today). Leibniz made the tactical error of accusing Newton of dishonesty. Newton always remained particularly paranoid about such accusations, living in secret terror that he would one day be accused of heresy over the Holy Trinity. On hearing of Leibniz' accusations, Newton became literally ill with rage. Even so, he honourably volunteered to let a committee of the Royal Society arbitrate on the matter. Leibniz agreed, seemingly not concerned that Newton was at the time president of the Royal Society. Newton then hijacked the committee's report and rewrote it in his own favour (though not in his own name). Leibniz continued bitterly disputing this verdict until his death in 1714. But Newton's rage, once provoked, was not so easily appeased. He continued to vilify Leibniz' name at every opportunity. Visitors spoke of him bursting into spontaneous tirades aganst the German philosopher, and Newton's subsequent scientific papers invariably included a furious irrelevant paragraph castigating his deceased adversary.

If this was how mathematics proceeded, what hope was there for science? Here chemistry stood at a heavy disadvantage. Physics, and physical properties, were easy to measure and describe in an exact fashion. Chemical changes and procedures were less amenable to precise description. With hindsight, we can now see that in many cases chemists were operating blind. They simply didn't know exactly what was happening in their experiments. All they knew was that something happened, and a product appeared.

This was very much the case with Brand and his method for producing phosphorus. He could have had no idea of the chemical constitution of urine. His initial intention had been to 'concentrate' the gold in the urine (by evaporation and distillation), and then 'reinforce' this by blending it with another golden substance long suspected by the alchemists of containing gold (i.e. sand). Phosphorus had been very much an accidental by-product of a sophisticated, if misguided, technique. Ironically, Leibniz played a part in this too. A few years after Krafft had purchased the secret from Brand, Leibniz turned up in Hamburg in the course of his work for the Duke of Hanover (father of George I). Leibniz prevailed upon Brand to put his secret at the disposal of his employer, and shipped Brand back to Hanover, with the intention of mass-producing phosphorus. Precisely what Leibniz had in mind for the quantities of phosphorus he hoped to produce is unclear. He had at one stage suggested it could be used to illuminate rooms at night, but had been told that this was impractical. (If the occupants of the room had not been poisoned by the choking fumes, they would certainly have been blinded.)

Undaunted, Leibniz secured large regular supplies of human urine for Brand from a nearby army camp, whose soldiers were renowned for the amount of beer they drank. Leibniz then heard that the workers in the mines of the Harz Mountains were even more renowned for the amounts of beer they consumed in the course of their hot and thirsty work. After obtaining permission from the bemused duke, Leibniz had extra supplies of human urine from the Harz Mountains transported in barrels by horse and cart over the 100 km of rutted roads to where the hapless Brand was set to work. Unfortunately, at this stage Leibniz was called away on urgent diplomatic business for his employer, and appears to have forgotten all about his phosphorus project. What became of Brand and his ever-increasing lake of human urine remains a malodorous mystery.

The secret of how to produce phosphorus would eventually be purchased in 1737 by the Académie Royale des Sciences in Paris, which immediately made it available to all scientists. Half a century later the Swedish chemist Karl Scheele found that phosphorus was a constituent of bones, and discovered a simpler, less distasteful method of extraction. Soon phosphorus was being used in the first matches, and in 1855 the Swedish manufacturer J. E. Lundström patented safety matches, which quickly made him a fortune. But this all came too late for Scheele to benefit from his discovery: he had died over half a century previously, in his forties.

Scheele was perhaps the most unlucky scientific discoverer of all time. During his comparatively short life he played a major part in the discovery of more elements than any other scientist before or since. Yet in the case of all the seven elements he discovered, his role was either eclipsed, disputed or overlooked.

Not that Scheele appeared to mind. He was a modest man, unsurpassed at his trade, that of the humble pharmacist. He was born in 1742 in Stralsund, on what is now the north-east coast of

Germany. During the early period of the eighteenth century this was still Swedish Pomerania, territory which had been conquered a century previously during the Thirty Years War. Despite his German name, Scheele looked upon himself as Swedish and wrote his scientific papers in that language.

Scheele was the seventh of eleven children. There being no money for his education, he was apprenticed to an apothecary at the age of fourteen. This was an inspired choice. Scheele's obsessive interest in the raw materials of chemistry meant that he never forgot a property, and his intuition concerning the constituents of chemical elements was unsurpassed. Scheele soon attracted attention and eventually obtained a post as an apothecary's assistant in the capital, Stockholm. This may sound small beer, but in 1775, at the age of thirty-two, he was elected to the Swedish Academy of Sciences – the first (and last) apothecary's assistant ever to achieve this high honour.

A short time later Scheele moved to the small provincial lakeside town of Köping (pronounced 'sherping'), in central Sweden. Here the local pharmacist had died, leaving the shop to his widow, Sara Pohl. Scheele bought the shop, and set up a laboratory, where he could pursue his experiments. Unbothered by the stifling provincial boredom of Köping, he buried himself in his work, while the widow Pohl and his sister kept house and took over the running of the shop. Occasionally distinguished Swedish and foreign chemists would visit, arousing much suspicion amongst the locals. Offers of professorships in Berlin, Stockholm and London arrived in the post, but were forgotten about. The supreme pharmacist felt no inclination to become a mere professor amongst professors.

For many years Scheele suffered from excruciating rheumatism and a series of other ailments. These were almost certainly brought on by his laboratory practices. Scheele was a firm believer in analysing first-hand the properties of the many substances he isolated and discovered. His laboratory notes even include an accurate description of the taste of hydrogen cyanide, which is extremely toxic and can cause a hideous and painful death even when inhaled or absorbed through the skin. (They also show that he could also have poisoned himself identifying by taste and smell the properties of hydrogen sulphide, a gas whose olfactory and lethal properties could be likened to those of a cross between a skunk and a rattlesnake.)

This was perhaps the most exciting period of chemical exploration. Freed from the straitjacket of alchemy, eighteenth-century chemists were able to reach out and explore the possibilities of an entirely new science. The undiscovered elements lay before them like the notes of a piano keyboard. They could play a few of these elementary notes, as well as a number of compound chords. But as their hands explored the keyboard, they gradually became aware for the first time of the vast range of tonal possibilities which lay before them. Scheele extended the possibilities of this keyboard more than any in his time. He identified a range of animal, vegetable and mineral acids, as well as discovering and identifying the contents of a large variety of important compounds.

But the intrepid Scheele survived, enabling him to turn his attention to the elements. In 1770 he became the first chemist to produce the gaseous element chlorine. Anyone who has visited a swimming pool will have experienced the eye-watering pungency of even very dilute chlorine: the effects on Scheele's one-man testing apparatus can only be imagined. Unfortunately, for once Scheele's chemical intuition let him down. He didn't recognize chlorine as an element, remaining convinced that the gas he had produced was a compound containing one of the gases in the air. Only thirty years later did the English chemist Humphry Davy (inventor of the miner's lamp) realize that this gas was an element. Indeed, it was Davy who first named it chlorine (from the Greek for 'light green'), on account of its appearance.

Davy became something of a nemesis for Scheele. The Englishman also trumped Scheele with the discovery of barium. Scheele did the important experimental work, distinguishing baryta (barium hydroxide), and again thirty years later Davy added the finishing touches by isolating the silvery-white metal barium, which received its name from the Greek word *barys*, meaning 'heavy'. Today barium is used in alloys to absorb impurities in vacuum tubes.

A distressingly similar sequence of events happened with the discovery of molybdenum, which was to become an important ingredient in the steel used for rifle barrels. Scheele had obtained a substance that he felt sure contained a new element, which could probably be isolated at a high temperature. As he had no furnace he passed the substance on to his friend the young chemist Peter Hjelm, who is now credited with the discovery of molybdenum. Around this time Scheele also succeeded in passing on to his friend the mineralogist Johann Gahn the secret which produced manganese, which mistakenly takes its name from *magnes*, the Latin word for a magnet, and has the property of dissolving water.

As can be seen from these discoveries, Sweden was developing an advanced and able community of chemists during these years. These pioneers were to play a major role in the development of chemistry. But this was far from being Sweden's only contribution. At the same time Linnaeus was laying the foundations of modern botany; the thermometer scale was established by Celsius; and, two centuries before Ford, the inventor Polhem conceived of the assembly-line factory. All this in a country of just two million inhabitants at the remote fringe of Europe. Comparatively speaking, Sweden's contribution to this period of the scientific revolution was the equal of any other country's. The eighteenth century was to prove Sweden's golden era. By the end of this period the Industrial Revolution was beginning to spread over Europe, and Sweden was providing over a third of the world's pig iron. The Swedish East India Company began trading as far afield as Japan; and, not to be outdone, the celebrated mystic writer Swedenborg produced several volumes describing in lurid detail his travels still further afield through heaven and hell. Even the king, Gustav III, collaborated on an opera. (The fact that he was later shot at the Stockholm Opera House had nothing to do with the critical reception of this work.)

Meanwhile, in the provincial wastes of Köping, Scheele continued with his thankless investigations of the chemical elements. By now Swedish chemistry was beginning to attract visitors from abroad. Two Spanish students, the brothers Don José and Don Fausto d'Elhuyar, called to see Scheele. In the course of their meeting Scheele explained how he had obtained from scheelite (which was named after him) a substance he called 'tungstic acid'. A year or so later the brothers d'Elhuyar succeeded in isolating from this the element tungsten, which means 'heavy stone' in Swedish. This element would one day be used in light filaments. The brothers d'Elhuyar later emigrated to America, where Don Fausto became Director General of Mines for Mexico.

Scheele's most important work was in the field which was to prove crucial to the next great advance in chemistry, namely, gases. Scheele managed to prove that air contained two distinct components, only one of which could support combustion. This he named 'fire air', and the other he named 'spoiled air'. From the latter he managed to isolate the element nitrogen, unaware that this had been done four years previously by the Scottish chemist Daniel Rutherford, uncle of the novelist Walter Scott. But it was Scheele's discovery of 'fire air' that was his greatest achievement. This was oxygen. As we now know, this element plays a major role in many of the most important naturally occurring chemical reactions. It was the element which held the key to the future of chemistry. Scheele first produced 'fire air' in 1772 by heating mercuric oxide, which readily gives up its oxgen and reverts to mercury. He included a description of this experiment in the only book he published, *Experiments in Fire and Air.* Due to a number of bungles, none of which were Scheele's fault, this work was delayed at the publishers and didn't finally appear until 1777. By this time the English chemist Joseph Priestley had published a description of the same experiment, thus scooping the most important elemental discovery yet made.

Unabashed, Scheele continued with his painstaking investigations, producing much original work which was far ahead of its time. Typical of this was his discovery of the effect light has on compounds containing silver. Half a century later the French artist and inventor Louis Daguerre would make use of this effect in the development of photography (which literally means 'writing by light'.)

But Scheele's first-hand involvement in chemical analysis eventually took its toll. In 1786, in his forties, he fell seriously ill. The symptoms suggest mercury poisoning. On his sickbed he married the widow Sara Pohl, so that she could re-inherit the apothecary shop, and within days he was dead. Scheele had freely corresponded with leading scientists throughout Europe, and the full extent of his intellectual generosity will perhaps never be known. Like his discoveries, his influence on the development of chemistry seems destined to be overlooked.

Two other important elements were discovered during this period.

German miners had long been deceived by an ore which much resembled copper ore. But unlike copper ore, which could tint glass blue when dissolved in acid, this ore tinted glass green. Consequently it was known by the superstitious miners as Kupfernickel: literally 'Old Nick's copper' (copper bewitched by the Devil, or false copper). In 1751 the Swedish mineralogist Axel

Cronstedt succeeded in isolating from kupfernickel ore a metal which bore no resemblance to copper. It was hard, silvery-white and was attracted by a magnet – a property not known in any other substance but iron. Cronstedt contracted the old miners' name, calling his discovery nickel. However, for many years scientists around Europe refused to accept nickel as a new element, maintaining it was a mixture of iron (hence the magnetic attraction) and other metals such as cobalt or copper (which, mixed with iron, was presumed to be responsible for the green effect). Only with the improvement of chemical analysis was Cronstedt proved right.

Cronstedt himself played a major role in this improvement. Previously minerals had been classified according to their physical properties – weight, colour, hardness and so forth. Cronstedt introduced the blowpipe into the analysis of minerals. This consisted of a long glass tube, which narrowed at one end. By blowing through the wide end it was possible to produce a narrow jet of concentrated air. When this was directed at a flame, it increased the heat of the flame and could focus this heated jet of flame on the object to be analysed. Cronstedt learned to distinguish the different colours the flame became when focused on a mineral. This enabled him to identify such important features as the gas arising from the ore, the colour of its oxide and the nature of the metals it contained. Using the blowpipe, Cronstedt carried out a systematic study of minerals, classifying them according to their chemical contents and chemical properties, thus becoming the father of modern mineralogy (literally, the study of substances that are mined). For the next century the blowpipe was to remain one of the key instruments for analysing the chemical contents of substances.

A century after Cronstedt discovered nickel, it was first used in the minting of coins by the Swiss. Seven years later, in 1857, the United States introduced nickel into the copper one cent piece. This was the original 'nickel'. Not until a quarter of a century later did the first five cent nickel coin appear, with one part nickel and three of copper: the modern US nickel. (Which may portray a president's head, but still derives its name from Old Nick.)

The other important element discovered during this period has one of the most picturesque histories of all the elements. In 1735 a French sailor walking along an estuary beach on the Pacific coast of Columbia came across some lumps of greyish clay the size and weight of cannon balls. Inside these he found deposits of a dull silvery metal. The French sailor brought several lumps of this metal back to his ship, where it was examined by a scientist who happened to be on board. This was the nineteen-year-old mathematical prodigy Don Antonio de Ulloa, who was participating in a project jointly sponsored by the Spanish and French governments. This project consisted of two expeditions – one sent to Lapland, the other to Ecuador – in order to measure the local degrees of meridian. These measurements were intended to help the Académie Royale des Sciences in Paris to determine the precise shape and dimensions of the earth.

On the return journey the French ship carrying de Ulloa put in at Louisberg, Cape Breton Island, off the Canadian coast. Here they discovered that the port had been captured by the British – who were now at war with France, but not with Spain. De Ulloa's papers, which included the secret of the shape of the earth and a description of a hitherto unknown metal, were impounded and sent to the Admiralty in London. De Ulloa himself, on the other hand, was treated with the courtesy and hospitality due to a visiting neutral gentleman, and given safe passage back to England. When he arrived in London he petitioned the Admiralty for the return of his papers. The Admiralty decided that the shape of the earth and a new metal more rare than gold were of no account, and returned de Ulloa's papers, which he then took home and published. The new metal was described as *platina del Pinto* ('little silver of the River Pinto'), and was reckoned to be of little commercial value. Unlike gold and silver, it was unmalleable and thus no use for ornaments. A few years later a solution to this problem was discovered by the Director General of Mines for Mexico, one Don Fausto d'Elhuyar. Writing to his brother Don José, who was now living in New Granada (Columbia), Don Fausto explained that when spongy deposits of the new metal were hammered together and then heavily compressed, the metal became as malleable as gold. The new *platina del Pinto* was soon being crafted into trinkets. More importantly, its resistance to chemical attack was found to be even greater than that of gold; as a result, it was soon being used in chemical apparatus.

The metal's original name, shortened to platina, was eventually altered by the English chemist Davy to platinum, to bring its feminine Latin name into agreement with those of other recently discovered metals such as barium and molybdenum. The idea of a feminine metal was evidently anathema to the Victorian English scientific establishment. This was to be the start of a distressing trend. All elements discovered since 1839, a couple of years after Queen Victoria's accession to the British throne, have been given the Latin neuter ending -ium, or the Greek neuter -on in the case of the inert gases. This sexless nomenclature was even extended to curium, which was named after Madame Curie. The sole exception to this rule is the element astatine, which has a feminine ending and is derived from the Greek *astatos*, meaning 'unstable'. This choice of gender was presumably made with no conscious derogatory intent, but one can't help feeling that it says something about the predominantly male society of chemists.

9

The Great Phlogiston Mystery

Chemistry was becoming full of possibilities. Its practitioners were now discovering an exciting range of new elements and compounds. This process was hastened by the development of new laboratory experiments. But most importantly, these latest ingredients and techniques were being used in a rational, methodical manner. Chemistry was no longer an embryo science. Yet for the time being its progress remained largely piecemeal, the result of uncoordinated experiments carried out by independent chemical pioneers. The developing science lacked a unifying principle. It also still contained certain puzzling features which defied explanation. One of these was fire.

The mystery of combustion had been the object of serious philosophical speculation since the ancient Greeks. For Heraclitus, fire had been the underlying principle of all substance and all change. Later Greek natural philosophers had suggested that all inflammable substances contained within them the element fire, which was seen as one of the four fundamental elements. This fiery element was released from a substance when it was subjected to appropriate conditions – such as heat, a spark or lightning, whereupon it manifested itself in flames. When the alchemists transformed the four elements into three (mercury, sulphur and salt), sulphur became the combustible element. Paracelsus explained how these three elements accounted for the burning of wood: the presence of sulphur in the wood enabled it to burn, the mercurial element provided the flame and the remains became ash because of the salt present.

Not until the latter half of the seventeenth century was a new explanation of fire proposed. The man who came up with the idea was one of the most remarkable frauds in the history of science. Johann Becher was born in 1635 in the little German town of Speyer, on the banks of the Rhine. The final chaotic years of the Thirty Years War deprived him of an education, and at the age of thirteen he set off across the ruined wasteland of Germany to seek his fortune. He travelled as far afield as Sweden and Italy, gathering a rough-and-ready education along the way. This included various alchemical ideas, and a smattering of business practice – but most of all the ability to present himself with supreme confidence, and an eye to the main chance.

In the aftermath of the Thirty Years War the map of Germany resembled nothing more than a fragile multicoloured vase which had been dropped from a great height. The German-speaking lands consisted of many scattered fragments of tiny states. In order to survive, each of these states had to control its economy and exploit its assets to the best of its ability. Industries such as brewing, textiles and ceramics mushroomed. And in order to supervise these microeconomies, each ruler required advisers and experts. At the age of twenty-six Becher managed to inveigle himself into the court of the Elector of Mainz as just such an expert. He then converted to Catholicism in order to marry the daughter of a rich and powerful imperial councillor, who bestowed on him a degree in medicine as a wedding present. On the strength of this, Becher managed to get himself appointed professor of medicine at the University of Mainz and personal physician to the Elector of Mainz. Soon after drawing his annual salary, he decided it was time for a move to Bavaria. Here he managed to talk his way into an even better paid post as chief adviser to the Elector of Bavaria, where he proceeded to put his knowledge of business into practice. He suggested that the way to economic prosperity for Bavaria lay in restricting trade with France (especially in silk), and setting up a local silk industry. Faced with ruin, the local merchants did their best to resist this plan, while Becher hastily constructed a silk factory. But he soon decided he was not sufficiently appreciated in Bavaria, withdrew his investment in the silk factory and moved on to Vienna. After boasting of his experience in the running of two states, he was taken on by the Emperor of Austria as chief economic adviser. His first move was to establish a large silk factory, which quickly bankrupted its Bavarian rival before running into financial difficulties of its own. He then suggested the digging of a canal linking the Danube and the Rhine to facilitate trade between Austria and Holland, proposed the adoption of a universal language (even producing a dictionary with ten thousand words), and outlined a project for transmuting the sands of the Danube into gold by alchemical means. But by now the emperor was beginning to tire of this cornucopia of fantastic schemes, and when the Danube sand project proved fruitless he had Becher escorted to the dungeons.

Unabashed, Becher turned up a year later in Holland, suggesting a scheme for the establishment of a silk manufacturing business in Haarlem. The dogged persistence with which he stuck to this silk-manufacturing obsession suggests that Becher genuinely believed it could be turned into a money-spinner. He also seems to have been equally convinced that transmutation into gold was possible. If not, it is difficult to account for the foolhardiness of his next move. Becher confidently presented himself before the Dutch Assembly, where he outlined an ambitious scheme for transmuting the vast tracts of sand along the Dutch coast into gold. The Dutch Assembly, which consisted largely of hard-headed businessmen, remained sceptical. Becher then demonstrated a preliminary test experiment, involving sand and a small amount of silver, which somehow succeeded in producing gold. As a result, the Dutch Assembly enthusiastically backed the scheme. Becher now explained that if he was to go into mass production he would need a considerable amount of silver to start off with, in order to produce the positively vast amounts of gold from the limitless supplies of sand. A few days before large-scale production was due to begin, Becher disappeared on the boat for London. Precisely how much of the silver vanished with him remains unclear, but the Dutch authorities continued to be keen on news of his whereabouts for some time.

Becher set up in London as a mining expert, a subject about which he had surprisingly acquired considerable expertise during his management of state economies. In this unaccustomed genuine role he even travelled to the mines of Cornwall and Scotland. But old habits die hard. On his return to London Becher wrote a treatise describing a clock that worked by perpetual motion, which he submitted to the Royal Society. To his surprise and disappointment, his consequent application to become a member of the society was turned down.

It is difficult to discern a scientific mind in the midst of all this, but Becher's works speak for themselves. In between his hectic business schedule he found time to read, and comment perceptively upon, the works of both his great chemical contemporaries – van Helmont and Boyle. In particular he noted that Boyle had not managed to produce a truly scientific theory of the elements. Boyle's ideas about indivisible elements and compounds may have been uncannily prescient, but they did in fact lack convincing experimental backing. (It is only with hindsight that we can recognize that his ideas were on the right track.)

Becher's masterpiece, *Physica Subterranea*, was published in Vienna in 1667 (during the lull between the silk fiasco and the Danube sand project). In this he advances his own theory of the elements, which was to have a profound effect on the chemistry of the next hundred years.

As is often the case with such contradictory characters, Becher reconciled himself to the irreconcilable aspects of his behaviour through a profound belief in God. The world had been created by God the chemist, who maintained his creation through a continuous process of alteration and transformation, change and exchange. (At one point he compares this to the workings of a well-run state economy, though he does not extend this analogy to include the silk business.) According to Becher all solid substances consist of three types of earth: terra fluida is the mercurial element, which contributes fluidity and volatility; terra lapida is the solidifying element, which produces fusibility or the binding quality; and the third element is terra pinguis (literally 'fatty earth'), which gives material substance its oily and combustible qualities. This is the principle of inflammability. Terra pinguis worked as follows: a piece of wood originally consists of ash and terra pinguis; when the wood is burnt, the terra pinguis is released, leaving the ash.

Becher's three elements are a recognizable development of the mercury, salt and sulphur theory proposed by Paracelsus and the alchemists. The notion of terra pinguis is not exactly original either. But the time was ripe for such an idea. At last a discernible principle had emerged to explain one of the main transformations which affected matter – burning.

Beside his belief in God, Becher also believed in chemistry. His affirmation of this belief is one of the most inspiring in all science: 'The chemists are a strange class of mortals, impelled by an almost maniacal impulse to seek their pleasures amongst smoke and vapour, soot and flames, poisons and poverty, yet amongst all these evils I seem to live so sweetly that I would rather die than change places with the King of Persia.'

The extent and consistency of Becher's commitment to this romantic belief is another matter. Johann Becher died penniless in London in 1682. He is said to have reconverted to Protestantism on his deathbed.

Becher's *Physica Subterranea* circulated widely throughout Europe, and went through several editions. By 1703 its third edition was being prepared in Germany by Georg Stahl, the professor of medicine at the newly founded University of Halle. Preceding Becher's text, Stahl inserted an introduction of his own. This expanded on Becher's original idea about terra pinguis, ensuring its central role in the development of chemistry during the eighteenth century.

Like his mentor Becher, Stahl was something of an oddity, though of an altogether different stripe. A crochety, misanthropic character, Stahl was a firm believer in Pietism, a puritanically inclined Protestant sect. Despite this, he married four times. This was not entirely due to his desire for new physical partners: his wives had a distressing habit of dying on him. His barren attitude towards society (fruitful, however, with regard to the science of the period) is encapsulated in his personal motto: 'In any dispute, whatever the general mass of opinion maintains is wrong.'

By the early eighteenth century Germany was entering a period of economic recovery, aided to a large extent by advances in the mining industry. Stahl had originally been drawn to Becher's *Physica Subterranea* on account of its central subject, mining (as indicated by the title). But when he read Becher's theory about the role of terra pinguis in combustion he quickly saw that this was only the germ of an idea which had not been fully developed. Stahl extended this idea into the field of mining.

The end process in the mining of metals was smelting. This involved heating the rocky ore with charcoal, so producing the molten metal. The technique of smelting had been known since prehistoric times, but what actually happened in the course of this process remained something of a mystery. Stahl recognized that smelting was now ripe for analysis from a chemical point of view, which could well lead to advances in mining technique. It was Becher's notion of terra pinguis which inspired Stahl to the understanding that smelting was simply the reverse process of combustion. In combustion, a substance such as wood released terra pinguis to become ash. In smelting, the ore absorbed terra pinguis from the charcoal to become metal. This insight was confirmed for Stahl by the fact that it also explained the process of rusting in metals. In rusting, the metal released its fiery terra pinguis and was reduced to an ash-like calx or rust. So rusting was simply combustion taking place at a slower rate.

The role of terra pinguis had now been extended beyond Becher's original conception, so Stahl renamed it 'phlogiston' – from the Greek *phlogios*, meaning 'fiery'. Stahl's phlogiston theory appeared to explain, in a fully scientific manner, several of the major mysteries of material transformation. Many chemists throughout Europe even began to suspect that phlogiston theory might hold the key to all chemical change. And it met every objection, too. Critics pointed out that neither combustion nor rusting took place without the presence of air. Why should air not be viewed as the essential ingredient in the process, rather than this mysterious phlogiston? Stahl recognized the difficulty, and half agreed. Yes, air was essential to the process – it served as the carrier of phlogiston from one substance to another.

More penetrating criticism came from Stahl's great rival Hermann Boerhaave, the professor of medicine at Leyden. In 1732 Boerhaave achieved fame and fortune by producing the first reliable textbook of modern chemistry, *Elementa Chemiae*. However, this was only written because a group of his students published a book of faultily transcribed notes from his lectures which became so popular that it threatened to ruin his reputation. And this was no small matter. In his day Boerhaave had a reputation throughout continental Europe on a par with Newton's. History may have revised this reputation, when it became clear that Boerhaave had in fact made no important original scientific discoveries, but his immense learning and powers of argument were not overestimated. This Stahl found to his cost, when Boerhaave opposed his phlogiston theory. If rusting was the same as combustion, why did no flame or heat accompany this process, Boerhaave demanded. This prompted Stahl to one of his most ingenious arguments. According to him, burning and rusting were the same process taking place at different speeds. When wood burned, the phlogiston escaped with such speed that it heated up the air which conveyed it and was rendered visible as flame.

But more damaging arguments were to come. There was no denying that combustible materials such as paper, wood and fat conformed to the phlogiston theory. Much of their substance disappeared upon burning, leaving only light traces of soot or ash. This was obviously due to the release of phlogiston. However, this was certainly not the case with the rusting of metals. Even the alchemists had noted that when a metal rusted away, the resultant accumulated rust weighed more than the original metal. And the reverse of this process only confirmed such evidence. Stahl himself had observed that when lead calx was heated and became lead, the lead actually weighed less than its calx, when according to phlogiston theory it had taken on phlogiston and thus should have weighed more. Even Stahl was nonplussed.

But the defenders of phlogiston theory were not slow to reply to such criticisms. (After all, the future rationale of chemistry was at stake here. Was this new science to be reduced once more to a mere jumble of accumulating data?) To Stahl's defenders, the answer was clear. There were evidently two types of phlogiston. The first sort – to be found in such substances as paper, fat, etc. – had weight. The second sort – as found in metals – did not. On the contrary, it had negative weight. This meant that when it attached itself to metal it actually buoyed the metal up, causing it to weigh less.

Others argued that there was in fact no problem here. Quite simply, the metal did not gain weight when it became a calx. This claim appeared contrary to much of the experimental evidence, including that obtained by Stahl. However, not all chemists agreed with this evidence. Ironically, this was due to improvements in laboratory practice, especially in heating methods. Chemists had now begun to use powerful magnifying glasses to concentrate the rays of the sun upon objects. These produced heat of such intensity that it often partly vaporized the metallic calx. This meant that the remaining calx would quite frequently weigh less than the original metal.

The calibre and ingenuity of the arguments over phlogiston theory indicate the sophisticated level to which chemistry had risen. This was now a fully-fledged science, with all its findings and theories open to rational debate. Strangely, at this point it didn't really matter which side was right, or which in the wrong. It was the debate itself which widened chemical understanding. (Curiously, the argument which appears most ludicrous – negative weight – has now achieved respectability in quantum physics.)

Stahl himself was not particularly concerned about the weight anomaly which appeared to disprove the existence of phlogiston. Indeed, many chemists of the period remained unconvinced by arguments based on measurement. Galileo had understood the importance of measurement in physics, and Boyle had stressed the need for experimental method in chemistry; but the importance of measurement in experiments was by no means universally recognized by chemists. Although chemistry had at last sloughed the motley skin of alchemy, it remained essentially a science of transformations. These were qualitative changes, and as such categorically different from the quantitative changes of physics. Measurement was not yet seen as central to chemistry.

Stahl was inclined to explain phlogiston as an immaterial principle, like heat. It simply flowed from one substance to another (carried by the surrounding air). In which case, questions of weight were irrelevant. This attitude accorded with Stahl's religious views. His quasi-mystic Pietist belief had led him to strongly oppose atheism and materialism, and adopt a vitalist philosophy: inanimate matter could receive life, or vitality, only from spirit. Part of Stahl's hidden scientific agenda was to discover how spirit interacted with matter. The phlogiston theory provided just such an explanation: phlogiston was the vital principle which animated matter with fire. The fact that this medievalist view underlay Stahl's scientific approach should not be allowed to detract from his science. After all, such mysticism apears positively down-to-earth compared with the religious beliefs of Newton, one of the few contemporaries who outshone Stahl. Yet even Newton's extreme views were compatible with science, after their own fashion. At the end of the seventeenth century, scientists still tended to be highly religious – and the central idea which united their science and their religion was highly seductive. In essence, it was identical to that of the early Muslim mathematicians. They believed that in uncovering the laws of science they were discovering how God's mind worked. God had created the world, and the laws of nature were nothing less than the way he thought.

Regardless of all objections, phlogiston theory was soon accepted by many progressive chemists. Quite simply, it explained too much for it to be abandoned. And in Stahl's view phlogiston explained much more than combustion. Here, he felt sure, lay the difference between acids and alkalis, perhaps the key to all chemical reactivity. It might even explain the colours and smells of plants. Such speculations appear far-fetched now, but Stahl was convinced there was a 'phlogiston cycle' in nature, and even cited experimental evidence to support this view. Wood certainly contained phlogiston, as was demonstrated when it burned. Likewise, van Helmont's celebrated experiment with the willow tree was more convincingly explained if the wood of the tree absorbed phlogiston from the air as it grew. With hindsight, it is easy to fault Stahl's line of argument – yet here in embryo is the idea of photosynthesis, the fundamental biological process which enables green plants to grow with the aid of carbon dioxide absorbed from the air. Not until fifty years after Stahl's death would this process be understood.

By this time the phlogiston theory was accepted by chemists throughout Europe. This was remarkable, since phlogiston had still never been isolated, or its presence even fleetingly detected in any experiment. Yet it was not altogether surprising. Van Helmont's suggestion that besides solids and liquids there might be a third state of matter called 'gas' was largely ignored. Chemists tended to focus their attention upon solids and liquids. When wood burned, leaving ash, the smoke and vapours were ignored. Even when it was confirmed that rust was heavier than metal, the possibility that this extra weight might have come from the air was not considered.

This state of affairs would not change until chemists carried out a systematic examination of the gases which made up air. The initial results of this proved spectacular. One of the first chemists to investigate the gases of the air quickly succeeded in isolating 'phlogiston'. This investigation was carried out by the English chemist Henry Cavendish.

In a country and an age which produced all manner of eccentrics, Cavendish may rightly be regarded as an eccentric's eccentric. His behaviour left the rest of the field baying and wittering far in the distance. He also happened to possess the finest scientific mind in England since Newton.

Cavendish was an aristocrat of the first water – descended on one side from the Dukes of Devonshire, on the other from the Dukes of Kent. His frail mother retired to the south of France to give birth to him, but survived only another two years. Following that curious trait of genius, young Henry was then brought up in a single-parent family. His father, Lord Charles Cavendish, was a keen and talented experimentalist (his work praised by Benjamin Franklin, no less), and he soon conscripted his already slightly odd offspring as his laboratory assistant. Young Henry had a peculiar squeaky voice, which could manage little more than a stutter – an affliction that remained throughout his life. Perhaps as a result, he developed an aversion to speaking with people, which soon developed into an aversion to any sort of human encounter. Men he tended to avoid, or simply ignore. Women he actively ran away from, covering his eyes – suggesting that during her brief two years of motherhood, Lady Cavendish had managed to stir certain Freudian perturbations.

When, at the age of forty, Cavendish inherited one of the largest fortunes in England, he immediately moved to a country house in unfashionable Clapham Common on the outskirts of London. Here he turned most of the rooms into extensive laboratories, and used a private side entrance so as to avoid visitors and servants. His housekeeper, who was given strict orders never to appear in his presence, was left her daily instructions in note form. He always dined alone, his meal laid out for him beneath silver covers on the table. When a director of the Bank of England had the temerity to call round and ask what he should do with the millions (Cavendish's account was at the time the largest deposit at the Bank of England), Cavendish dismissed him with a flea in his ear. The banker was told never again to plague him with his presence and just get on with his job.

To his further irritation, Cavendish had now begun to attract attention when he went walking in the street. This was hardly surprising, since he insisted on dressing in threadbare family

Henry Cavendish

hand-me-downs, many of which had gone out of fashion in the previous century.

Cavendish had shown exceptional promise from the start, and his scientific endeavours had soon begun to yield fruit. As a result of just a handful of published papers he was elected to the Royal Society at the singular age of twenty-nine. Surprisingly, he became a regular attender at the society's meetings, a habit he was to retain for the rest of his life. Yet as one of his fellow members observed, he 'publicly uttered fewer words in the course of his life than any man who ever lived to fourscore years, not at all excepting the monks of La Trappe'. But this Trappist reticence does not mean that he was entirely silent. According to another fellow member, he was in the habit of giving a 'shrill cry . . . as he shuffled quietly from room to room'.

Curious though it may seem, this behaviour is evidence of Cavendish's total dedication to science. He was willing to suffer these social gatherings, which evidently caused him such anguish, for the sake of keeping abreast of scientific developments. He was a great believer in the public sharing of the latest discoveries, for the good of science as a whole. Yet for one who spent the bulk of his waking life in his laboratory, Cavendish's published papers were comparatively few and far between. Fortunately one of these included his pioneering work on gases.

Cavendish became intrigued by the gas which was produced when certain acids reacted with metals. This gas had earlier been isolated by Boyle, and more recently by Cavendish's contemporary Joseph Priestley; but as Cavendish was the first to investigate its properties in a comprehensive scientific fashion, its discovery is usually associated with him. He pioneered the weighing of particular volumes of different gases in order to discover their different densities. Initially he did this by weighing a bladder of known capacity filled with different gases. Later he invented his own ingenious apparatus, which ensured greater purity of the gases to be weighed. Cavendish discovered that this new gas, produced when certain acids reacted with metals, had a density of only one fourteenth that of air. He also observed that when a flame was introduced to a mixture of this gas and air, the gas caught fire. He thus named it 'inflammable air from the metals'. Cavendish mistakenly thought that the 'inflammable air' actually came from the metals, rather than from the acid. Like most of his contemporary chemists, he also subscribed to phlogiston theory, believing that metals were a combination of calx and phlogiston. This, together with the exceptional lightness and inflammability of 'inflammable air', led him to the sensational conclusion that he had managed to isolate phlogiston!

But this one step backwards was accompanied by a significant step forwards. Priestley had noticed that when he ignited a mixture of air and 'inflammable air' in a flask, moisture was left behind on the glass of the flask. Cavendish seized on this hint, produced larger quantities of the moisture and discovered that it was water. This meant water was in fact a combination of two gases – air and 'inflammable air'. Water could no longer be viewed as an element: this was the final nail in the coffin of the Aristotelian four-elements theory.

Around the same time (1784–5) this same experiment was also carried out by James Watt, the developer of the steam engine. This led to an unseemly squabble over priority, into which Cavendish and Priestley were reluctantly drawn. Discoveries were now increasingly being made simultaneously. (Watt's development of the steam engine, for instance, was contemporaneous with several similar projects.)

Science was developing into a body of knowledge which frequently prompted those working within it in the same direction. This has led to the understanding that scientific discovery is to a certain extent predetermined. If 'inflammable air' had not been discovered by Cavendish (or Boyle, or whoever), it would sooner or later have been discovered by someone. Science could now be viewed as a cultural-historical phenomenon, rather than simply the creation of individual geniuses working alone.

But it was neither Cavendish nor Priestley nor Watt who was to give this newly discovered gas its modern name. This happened a decade later when the great French chemist Lavoisier called it 'hydrogen' (from the Greek *hydro*, 'water', and *-gen*, 'generator, begetter, maker'.)

Cavendish made several important discoveries concerning the compostion of gases (he even suggested the existence of an inert gas a century before this class of gases was discovered). He also made electrical experiments which were far ahead of their time. (With no method yet devised for measuring electrical current, he would subject himself to a shock and estimate the pain!) On top of this he produced a veritable cornucopia of experimental results – most of which remained unpublished in his laboratory notebooks. Yet this withholding of information was not some perverse eccentricity. Cavendish set himself the highest standards, refusing to publish any work that did not satisfy them, or which continued to puzzle him.

Perhaps his greatest experiment was to weigh the world (no less!). Here the exceptional calibre of his work is plain to see. Newton's formula for gravitational attraction included the weights of the two objects concerned, the distance between them and a gravitational constant G, whose value remained unknown. (Newton's formula thus gave a proportional relation, rather than precise figures.) Cavendish conceived of a brilliant experiment for discovering the value of G. He attached a lead ball to one end of a horizontal rod, and another to its other end. The mid-point of the horizontal rod was suspended by a thin wire so that it could swing freely. In this way he was able to study how the freely suspended lead balls were affected by the gravitational pull of two larger stationary lead balls. Such was the precision required in the minute measurements involved that Cavendish's experiment was later recognized as 'the beginning of a new era in the measurement of small forces'. Cavendish was the first to experiment with the minute gravitational attraction of small objects. Knowing the weight of the different lead balls and the distances between them, and having measured the gravitational pull between them, Cavendish then knew all the figures in Newton's formula, and was able to calculate the value of G. He could now substitute this in the equation giving the attraction between the earth and an object on its surface (whose small weight was easily measured). He was thus able to calculate the unknown weight of the earth. This he found to be 66×10^{20} tons – a figure of such accuracy that it was not improved upon until the advent of the twentieth century. It also brought to an important conclusion work which Newton had initiated a century earlier.

After seventy-nine years of healthy solitude Cavendish fell ill, and his condition quickly deteriorated. His manner of living remained true to the end. According to the contemporary historian of chemistry Thomas Thomson:

> When he found himself dying, he gave directions to his servant to leave him alone, and not to return till a certain time which he specified, and by which period he expected to be no longer alive. His servant, however, who was aware of the state of his master, and was anxious about him, opened the door of the room before the time specified, and approached the bed to take a look at the dying man. Mr Cavendish, who was still sensible, was offended at the intrusion, and ordered him out of the room with a voice of displeasure, commanding him not by any means to return till the time specified. When he did come back at that time, he found his master dead.

Cavendish left none of his millions to science. Contrary to popular belief, the Cavendish Laboratory at Cambridge, founded sixty-one years after his death in 1871, was in fact financed by a relative. But in this way the name Cavendish was to remain associated with scientific excellence. The laboratory also remained associated with Cavendish himself, and not only through a popular misconception. The first director of the Cavendish Laboratory, the great pioneer of electromagnetism James Maxwell, collected Cavendish's unpublished papers and even re-created several of his electrical experiments. This showed that Cavendish had anticipated much of the work of the two great pioneers in the field, Faraday and Maxwell himself. These experiments coincided with the first period of epoch-making scientific work undertaken at the Cavendish. (The second came a generation later with Rutherford's ground-breaking research into the structure of the atom, and the third was the discovery of the structure of DNA by Crick and Watson in 1953. No other laboratory in the world has yet been able to match such achievements.)

Cavendish had apparently discovered phlogiston, thus supporting the 'three earths' elemental theory proposed by Becher. Yet this theory was essentially a derivation of the Aristotelian four-elements theory, which appeared to have been exploded by Cavendish's discovery that water consisted of two separate gases. So what exactly was phlogiston? Could this be the key to the true nature of the elements?

The answer obviously lay in the further investigation of gases and combustion. This topic was also beginning to attract attention for more practical reasons. Watt's invention of the improved steam engine had introduced a new concept of energy. For the first time in history humanity was no longer dependent upon its own muscles or those of animals (or the vagaries of wind or watermills) as a source of power. Potentially unlimited energy could now be summoned at will. The machine age was beginning, and already in Britain this was giving rise to an industrial revolution. Ill-paid rural labourers began leaving the land to work in factories, often in appalling conditions. (In the poet Blake's vision, these would become the 'dark satanic mills' disfiguring 'England's green and pleasant land'.) The resultant social disruption brought about discontent and a questioning of traditional values. One symptom of this was the spread of dissenting religious sects, such the Unitarians, whose extreme form of Protestantism placed them beyond the pale of the established Protestant Church of England. Amongst the most influential leaders of the Unitarian movement was Joseph Priestley, one of the few contemporary scientists with whom Cavendish maintained a sporadic correspondence.

Priestley was born in 1733, just two years after Cavendish. His mother too died when he was a child and, like Cavendish, he suffered throughout his life from a hampering stutter. Yet in other aspects they were opposites. Priestley cared deeply about the plight of humanity. Like Newton before him, he considered his religious work to be of more importance than his scientific work. And in this case, such self-assessment was probably correct. Priestley's experiments may have changed the course of chemistry, but his religion-inspired compassion for his fellow men had the nation up in arms.

Priestley was first and foremost a Unitarian minister and, to begin with, practised science only as a hobby. In 1765 his sympathies with the oppressed American colonists brought him into contact with the American patriot Benjamin Franklin, another excellent scientist with strong religious beliefs. Yet, paradoxically, it was Franklin's example that persuaded Priestley to take his science more seriously.

Two years later Priestley took over a ministry in Leeds, where he found himself living next to a large brewery. Here he began a systematic investigation of gases. The bubbling vats of fermenting beer next door provided an ample supply of the gas then known as 'fixed air' (carbon dioxide). Priestley collected the gas and observed its properties. A mouse placed in it soon died; likewise, a flame went out; and when weighed, the gas was found to be heavier than air. In order to obtain a more pure sample of 'fixed air', Priestley resorted to the time-honoured method of adding chalk to a flask of sulphuric acid. The gas was allowed to escape from the sealed effervescing flask through a tube which looped beneath the surface of a container of water and up into an overturned, water-filled jar, so that the gas bubbled up into the jar, displacing the water. Here Priestley noticed another characteristic of 'fixed air'. Part of the gas appeared to dissolve in the water. The result was 'a glass of exceedingly pleasant sparkling water'. Priestley had discovered what we know as soda water. (Even before the birth of the nation, an American sympathizer had invented the process which would one day give the world Coca-Cola.)

Priestley now followed a tip from Cavendish, and began collecting his gases over mercury instead of water. This enabled him to discover, and investigate, a number of new gases which had not previously been collectable, as they dissolved in water. In this way he isolated the gases which we now know as nitrogen oxide, ammonia, hydrogen chloride and sulphur dioxide. None of these gases appeared to be elements, and all except ammonia became acidic when dissolved in water.

It was Priestley's use of mercury which led in 1774 to his most important discovery. When heated in air, mercury forms a brick-red calx. Priestley placed a sample of this in a flask. He then subjected the sample to concentrated heat by focusing the sun's rays on it through a magnifying glass 1 ft in diameter, which he had just acquired (and was dying to use). He found that silvery globules of mercury began appearing in amongst the red calx. Simultaneously the decomposing calx emitted a gas. When

Priestley collected this gas he found it appeared to be 'a superior air', which possessed some remarkable properties.

> A candle burned in this air with an amazing strength of flame; and a bit of red-hot wood crackled and burned with a prodigious rapidity, exhibiting an appearance something like that of iron glowing with a white heat, and throwing out sparks in all directions. But to complete the proof of the superior quality of this air, I introduced a mouse into it; and in a quantity in which, had it been common air, it would have died in about a quarter of an hour, it lived . . . a whole hour.

Indeed, it might have lived even longer if Priestley hadn't accidentally allowed the mouse to die of cold. He even tried sampling this new air: 'The feeling of it to my lungs was not sensibly different from that of common air; but I fancied that my breast felt peculiarly light and easy for some time afterwards.' He became high! As a result, he forecast that this gas might have a future as a fashionable vice amongst the idle rich, a sort of mild precursor of cocaine.

Priestley explained the chemistry of his newly discovered drug in terms of phlogiston theory, in which he was a firm believer. Because objects burned so easily in its presence, it meant they rapidly released their phlogiston. The explanation of this phenomenon seemed clear. The new gas was a form of air devoid of phlogiston, which was why it absorbed phlogiston so rapidly. For this reason, Priestley named his new gas 'dephlogisticated air'.

Priestley was a great believer in scientific openness, and when he visited Paris later in 1774 he passed on his discovery of 'dephlogisticated air' to the greatest chemist of the era, Antoine Lavoisier. This was to have a far-reaching effect on the history of chemistry.

Indeed, this is the reason why Priestley is still so often

accredited with the discovery of oxygen, though it is now known that Scheele had in fact performed precisely the same experiment with mercury calx two years previously in Sweden. Priestley's discovery may not have been strictly original (except for him), but it was he who passed it into the scientific domain, where it was to have great effect.

Such allocation of priority is not without precedent. A similar course was to be adopted with regard to several original discoveries made by the nineteenth-century German mathematician Karl Gauss, who along with Archimedes and Newton is generally regarded as one of the three supreme mathematicians of all time. Gauss kept some of his more daringly original work secret in unpublished notebooks. Thus it was not known that he had made major advances in number theory, discovered non-Euclidean geometry while still in his youth, founded the theory of knots, whose significance is only now being fully understood, and even invented the telegraph. Almost all of these discoveries have been credited to others who discovered them independently later. But it was they who passed them on into the public domain, where they could contribute to further advances in their field. Discoveries are for all, or for nothing. (Even those kept secret for commercial or military purposes at least enter the public domain in effect.)

Priestley's insatiable curiosity and belief in experiment led him to make, and make public, a huge variety of chemical discoveries. His contemporary Davy even claimed, 'No single person ever discovered so many new and curious substances.' Priestley's fame spread far and wide. His soda water, which he refused to patent, became a health craze throughout Europe (though its adoption by the British navy as a cure for scurvy proved a failure). Another lasting memorial is his naming of the sap from a Brazilian plant which had recently arrived in Europe. When he found that this sap rubbed out pencil marks, he called it 'rubber'.

Yet his scientific renown was no match for his political and religious fame. Though this was less welcome. In 1776 he applauded the declaration of independence by the American colonists. The French Revolution in 1789 evoked a similar response. At last, he hoped, there might be just societies in this world. Anti-French feeling ran high, and two years later Priestley's home in Birmingham was burned down by a 'King and Country' mob incited by the authorities. Priestley escaped to live in London, where he was shunned. In recognition of his support for the revolution, he was made an honorary citizen of France. This counted heavily against him when France declared itself a republic, guillotined the deposed Louis XVI and finally declared war on Britain. Priestley decided to emigrate to the United States. In April 1794, at the age of sixty-one, he packed up his chemical apparatus and set sail for New York. To the end, he retained the characteristic tolerance and optimism encouraged by his rational mind and religious beliefs. 'I do not pretend to leave this country . . . without regret,' he declared, but 'when the time for reflection shall come, my countrymen, I am confident, will do me justice'.

When Priestley sailed for America, he was still convinced that phlogiston explained the problem of combustion. Others had already begun to question this.

10

The Mystery Solved

When Priestley demonstrated his 'dephlogisticated air' to Lavoisier in Paris, Lavoisier immediately grasped the far-reaching significance of this discovery. Priestley may have been a superb experimentalist, but Lavoisier had the superior theoretical understanding of chemistry. As no other, he used his encyclopedic knowledge of chemistry to establish a scientific structure for this field. Only now, with the plethora of new discoveries, was this becoming possible. Patterns were beginning to emerge: resemblances which pointed to groups of related substances, their similar properties possibly caused by similar types of elements.

Chemistry was now ripe for a supreme scientific mind.

Not for nothing has Lavoisier become known as the Newton of chemistry. Yet he was no single-minded pioneer. During his brief fifty years of life he not only established modern chemistry, but also found time to occupy (simultaneously) several top-level administrative positions, as well as contributing technological advances in a number of disparate fields: ballooning, the mineralogical mapping of France, urban street lighting, the Paris water supply, the efficiency of gunpowder and a full-scale model farm, to name but a few. Yet who was this French colossus? The historian of science Charles C. Gillespie famously characterized Lavoisier as 'the spirit of accountancy raised to genius'. And there is some justification in this. For someone who accomplished so much, in so many fields, who lived a life of such immense variety during one of the most exciting periods of French history, there remains surprisingly little evidence of personality. The man who did everything was nothing. He was simply Lavoisier.

Antoine Lavoisier was born in 1743, into a wealthy upper-middle-class family. As he was growing up the first volumes of Diderot's *Encyclopédie* began to appear. This was the ultimate product of the French Enlightenment, which had started with the likes of Descartes. It was intended as a 'rational dictionary' to propagate the widespread flowering of the arts and sciences that was now taking place throughout Europe. Its contributors included the cream of France's eighteenth-century intellectuals: Voltaire, Rousseau, Montesquieu, the mathematician d'Alembert, the philosopher Condillac, Diderot himself. Interestingly, it was Diderot who insisted that the future of science now lay in experiment, not mathematics. The vision of Newton was giving way to the practical approach of Cavendish and Priestley – an insight which was to have a profound influence on the young Lavoisier.

The *Encyclopédie* was intended to combat the stifling repression of the Church and the *ancien régime* – epitomized by Louis XV and the ruinous extravagance of his vast court at Versailles (with over ten thousand rooms and closets, but no water closets). Yet Louis' celebrated mistress Madame de Pompadour was not referring to the plumbing when she declared with insouciance: '*Après nous le déluge*' ('After us the deluge'). Unless enlightenment spread and there were sweeping reforms, France was heading for a major social catastrophe.

Lavoisier showed such exceptional talent that at the age of twenty-three he was elected to the Académie des Sciences (the former Académie Royale des Sciences). His next move came as something of a shock to his enlightened fellow Academicians. He chose to join (and invest his inheritance in) one of the most oppressive agencies operating under the *ancien régime*, the Ferme Générale. This was the privately run organization which collected certain government taxes on a commission basis. As a result, the vicious methods used by the profit-driven tax collectors of the Ferme Générale made this one of the most loathed institutions in a land where all forms of authority were becoming increasingly despised. Certainly Lavoisier was no collector – he took only an administrative post in the Ferme – but he must have been well aware of what was going on. The excuse usually given is that he wished his investment to earn sufficient income for him to continue with his scientific researches.

But Lavoisier's first and greatest benefit from this employment was not to be financial. At the age of twenty-nine he unexpectedly married the thirteen-year-old daughter of one of the senior partners of the Ferme. His bride, Anne-Marie, is said to have been hastily married off by her father to avoid royal pressure for her to marry a fifty-year-old penniless aristocrat – whom the bewildered young girl described to her father as *une espèce d'ogre* ('a sort of monster'). Despite the circumstances, this hurried marriage by such an elegible bachelor to a thirteen-year-old would seem to suggest either weakness of character on Lavoisier's part or naked careerism. To say the least. Whichever way, such reservations pale into insignificance in the light of the marriage itself. Though this was childless, Antoine and Anne-Marie Lavoisier were to remain for the rest of their lives a deeply devoted couple. It was a marriage that made both their lives. Anne-Marie was to become her husband's partner in science. Here was more than just a collaborator. She learned English so that she could inform him of the latest chemical papers delivered at the Royal Society. She also collaborated with him on many of his most important experiments – which were sometimes installed by her, and invariably written up by her. And she appears to have contributed considerably during their discussions of chemical theory. (Parallels have even been suggested with the partnership between Pierre and Marie Curie at the turn of the following century – but this would seem to overstate the case.)

Yet one thing is certain. Without Anne-Marie's support, Lavoisier would certainly have had neither the time nor the energy to pursue his successful administrative career. It was this which enabled him to accumulate the fortune that financed his scientific and humanitarian schemes. Amongst the latter was the model farm he established on his land in provincial Orléans, and the organization for famine relief which he set up throughout the same region. At least some of what he extracted in tax was returned in charity.

The abilities Lavoisier exhibited in his work for the Ferme soon saw him promoted to a senior partner. At the Académie des Sciences he began to be appointed to various committees. Likewise, his suggested improvements for the munitions industry eventually secured his appointment as Director of the Gunpowder Administration. This last post included a grand residence at the Arsenal in Paris, where he installed a superbly equipped laboratory.

It was here that Lavoisier and his wife sat for the celebrated portrait by David, the finest artist of the age. Despite its formality as a society portrait of 'the great scientist and his wife', this is also a portrait of single-minded, if somewhat unemotional, devotion. Lavoisier is sitting writing at his red velveteen-covered table, on which stand various pieces of glass apparatus. Anne-Marie is leaning over him, her hand resting lightly on his shoulder, while his half-turned face gazes up at hers. He is elegantly bewigged, outfitted in fashionable lace choker and matching cuffs, elegantly cut black knee-breeches and matching stockings; while she is dressed in a full-length, full-blown blue beribboned taffeta salon dress. The pieces of apparatus appear glossily precise and pristine, and one feels that even the large fragile flask lying on its side by his buckled shoe is in no danger of being shattered. This is a portrait of the scientist as a man of reason, conducting

David's celebrated portrait of Antoine Laurent de Lavoisier and his wife, painted in 1788

his precisely measured experiments in a clean, well-lit laboratory, assisted by his respectably attired wife. It can't always have been like this; there must have been days of frustration and acrid fumes, with heads in hands and cracked tubes littering the bench tops. But gone for ever are the days of the solitary alchemist lurking amidst the choking fumes of his dim den. Chemistry is now a civilized science.

According to Anne-Marie, Lavoisier managed to achieve so much in his life by adhering to a strict routine. Each morning he would rise at six, working till eight on his science. The ensuing daylight hours would be divided between business at the Ferme Générale, the Gunpowder Administration and committees of the Académie des Sciences. At seven in the evening he would retire for another three hours of scientific studies. Sunday, his *jour de bonheur* ('day of happiness,' as he called it), was set aside for conducting experiments. However, this routine cannot have been that inflexible, as the Lavoisiers were famous for the dinners they gave at their residence in the Arsenal. These were attended by leading scientists and intellectuals (including on occasion Thomas Jefferson and Benjamin Franklin). After coffee and cognac, the guests would be conducted to the laboratory for a demonstration of Lavoisier's latest findings.

From the outset, Lavoisier adopted a modern approach to chemistry. This was epitomized by his belief in the balance (such scales were the most precise weighing apparatus available at the time). Chemistry was nothing to do with mysterious transformations: all change could be explained and could also be measured. At the same time he felt that chemistry was being severely hampered by traditional theories which in fact explained nothing and even appeared to be preventing further breakthrough. In the light of this, it comes as something of a surprise that Lavoisier was initially inclined to believe in the four-element theory. He had been persuaded to this view by experimental evidence. With his own eyes he had seen how water, when boiled for hours on end in a flask, eventually produced a tiny sediment. Water appeared to contain earth, and could also boil away into air. But he still retained niggling doubts. The fact was the four-elements theory explained nothing about chemical behaviour and just led nowhere.

In 1770 Lavoisier resolved to put this apparent transformation of water into earth to an exhaustive test, under the strictest scientific conditions. For this he used a 'pelican', a sealed container where water could be boiled. Within the pelican the water vapour passed into a tube where it was condensed, and the condensed water then returned to the boiling flask, so that nothing was lost in the process. Before starting the experiment, Lavoisier separately weighed the water and the pelican. The water was then boiled in the sealed pelican for no less than 101 days. As before, a sediment duly appeared. Lavoisier then weighed the water, and found that it weighed precisely the same as it had done previously! The sediment could not have come from the water. When he weighed the pelican, he found that it weighed slightly less than before – and the difference was exactly the same as the weight of the sediment. The 'earth' had not come from the water, it had been extracted from the glass by the boiling water. The last possible evidence for the four-elements theory collapsed.

Two years later Lavoisier turned his attention to the vexed problem of combustion. He conducted an experiment heating lead in a sealed vessel containing a limited supply of air. To begin with, the surface of the lead formed a layer of calx, and then it stopped doing so. According to orthodox phlogiston theory, the lead had released its phlogiston to become calx, until the air in the container had absorbed as much phlogiston as it could; then the process had stopped, because the air was saturated with phlogiston.

Lavoisier now weighed the entire apparatus (containing lead, calx, air, etc.), and found that it weighed exactly the same as it had done before being heated. Next he weighed the lead and its coating of calx – and found, as others had found before him, that this weighed more than the metal had previously weighed. But if the metal had gained weight when it was partly turned into calx, then something else in the vessel must have lost a similar amount of weight. This could only have been the air. But if the air had diminished, this would have meant there was a partial vacuum in the vessel at the end of the experiment. Lavoisier repeated the experiment, and found that when he opened the sealed vessel there was a tiny hiss of air rushing into the vessel. It had indeed contained a partial vacuum!

Lavoisier's experiment proved that when a metal turned into its calx, this had nothing to do with the loss of some mysterious phlogiston (which had negative weight, or alternatively was some immaterial 'principle' and thus had nothing to do with loss or gain of weight). He had shown that the metal in fact combined with a material substance which had weight, and that material substance consisted of a portion of the air.

It was now that Priestley arrived in Paris, and demonstrated to Lavoisier the new elemental gas he had discovered, which he called 'dephlogisticated air'. Lavoisier regarded this Nonconformist English minister as little more than an amateur, a well-meaning fellow of admirable liberal principles, but hardly a scientist of the first rank. Admittedly he was an experimentalist of some talent, but he lacked any sound theoretical understanding of science. Like so many Englishmen he was a pragmatist, rather than a man of reason – no match for a French theorist. Priestley may have stumbled across this new gas, but he had no idea of the significance of what he had discovered. Lavoisier understood this at once. How could this gas be 'dephlogisticated air', when he had already proved that there was no such thing as phlogiston.

As soon as Priestley returned to England, Lavoisier repeated Priestley's experiment and obtained so-called 'dephlogisticated air'. He then proceeded with some more sophisticated experiments, and found that 'dephlogisticated air' was present in all air. Lavoisier carried out an experiment with a burning candle. This was placed on a float in a bowl of water. Lavoisier then placed an upturned glass jar over the candle, with the rim of the jar beneath the surface of the water. As the candle burned, the water gradually rose within the beaker – the candle was using up some of the air. But Lavoisier noticed that the candle always went out when the water had risen to occupy one fifth of the flask. Air obviously consisted of two gases, in proportions of one to four. The one fifth which was used in combustion was Priestley's so-called 'dephlogisticated air'. Lavoisier now realized that what actually took place during combustion was very much the opposite of phlogiston theory. When something burned, it didn't release some mythical phlogiston, but rather combined with the so-called 'dephlogisticated air', which made up one fifth of the air.

Lavoisier decided to rename this element 'oxygen' – from the Greek *oxy-* meaning 'acid' and *-gen* 'generator or producer'. This was a rational name for the new gas: Lavoisier's experiments had led him to the conclusion that this elemental gas was present in all acids. (This appeared to be true at the time. It was only disproved when Davy discovered chlorine was an element, a generation after Scheele had originally discovered this pungent green gas. Davy went on to show that hydrochloric acid contained hydrogen and chlorine, but no oxygen, and the name oxygen was thus shown to be anomalous. But by then it was too late to change it. So oxygen joined the great universal misnomers, along with the West Indies, the Cultural Revolution and the like.)

At last the phlogiston theory was finally exploded. Lavoisier published a paper describing the role of oxygen in the process of combustion. Deviously, he omitted any mention of the vital role Priestley's work had played in this momentous discovery – and even went so far as to imply that he, Lavoisier, had really discovered oxygen. (Lavoisier knew he was a great chemist, but longed to capture the public imagination by discovering an element.) It is said that Madame Lavoisier celebrated the publication of her husband's paper with a 'rational scientific ceremony'. A hundred years after the death of Becher, the original begetter of the phlogiston theory, Madame Lavoisier dressed herself in the robes of an ancient Greek priestess and ceremonially burned the works of Becher and Stahl on an altar before a gathering of scientific luminaries.

Despite this dramatic gesture, not all were fully convinced. With the advent of the Industrial Revolution, science had now emerged as a social phenomenon, and as such was becoming dragged into European nationalistic rivalries. (Determined to maintain their industrial lead, the British attempted to block the export of any new machinery, manufacturing technology or even skilled workers until well into the nineteenth century.) For the Germans, phlogiston theory was the work of their great chemist Stahl. What did the French know about science? Lavoisier couldn't possibly be right. Meanwhile the English remained as aloof as ever from such continental bickering: both Priestley and Cavendish obstinately clung to the phlogiston theory their work had done so much to confirm.

Combustion was recognized as one of the main chemical processes; its similarity to rusting had already been noticed by Stahl. Lavoisier now conducted a series of experiments which demonstrated that respiration too was a similar process. Air inhaled contained a much higher proportion of oxygen than air exhaled. During respiration oxygen was inhaled and 'fixed air' (carbon dioxide) exhaled. There are several drawings of Lavoisier conducting these famous experiments on respiration using a human guinea pig. The subject of the experiment is seated in a chair stripped to the waist, pedalling a machine. A gruesome mask is fitted closely over his face, obscuring the entire front of his head so that it appears like that on a tailor's dummy. A tube leads from the front of the mask through various flasks – while Lavoisier directs his bewigged, frock-coated assistants, and Madame Lavoisier sits watching intently at her desk, taking notes.

As in all of Lavoisier's experiments, everything was precisely measured. The subject was experimented upon at room temperatures ranging between 77°F and 54°F. He was measured 'with food', 'without food', 'at work without food' and so on. Lavoisier operated on the principle that the substances taking part in a chemical reaction could be transformed, but their overall weight would always remain the same. This had been the secret of his discovery that water boiling in a flask does not produce earth. It is this idea which is behind the law of conservation of matter. This basic assumption of Lavoisier's experiments was to become one of the cornerstones of nineteenth-century chemistry.

Lavoisier's laboratory

As well as indicating chemistry's future, Lavoisier had also solved one of the great enigmas of its past, combustion, and shown that this solved much more than the mystery of fire. Combustion was oxidation, the addition of oxygen – which in burning formed ash, in rusting formed 'calx', in respiration formed 'fixed air' (carbon dioxide) and so forth.

As can be seen from the above names, chemistry was now stumbling into a morass of contradictory nomenclature. Scientifically formed names such as oxygen (even if anomalous) were being used alongside theoretically speculative names such as 'fixed air', and remnant alchemical names such as calx – which in Latin meant lime! Likewise 'dephlogisticated air' and oxygen now began to appear in separate textbooks, each naming the same substance, but according to different theories. And these were just the names of simple gases. When it came to more complicated substances and compounds, there was an even more baffling legacy of nomenclature – some names stemming from nature and mining terms, others from effects or properties (real or imagined), and still more from alchemical practices. Different languages had different names for substances – and different specializations, such as medicine or geology, often had their own terms too. There was tincture of this, oil of that, essence of whatever; quicksilver and hydrargyrum were two of many names for mercury; litharge was what we now know as lead oxide; alum was the popular name for aluminium potassium sulphate. The last two substances provide the clue to the way forward. In the book which Lavoisier co-authored in 1787, *Method of Chemical Nomenclature*, a logical system of chemical naming was proposed. In future all compounds should be named rationally, after the elements of which they were composed. For instance, the name hydrochloric acid would indicate that this substance was a compound of hydrogen and chlorine; such nomenclature would render the chemical composition transparent. And the effect would transform our reading of what happened in chemical reactions. For example, we would know that when zinc is added to hydrochloric acid, the resulting chemical reaction forms the compound zinc chloride – and by logical deduction, the gas given off by this effervescent reaction would obviously be hydrogen. Chemistry was making progress through measured, scientifically conducted experiments – and these could now be described in scientific language.

It is almost impossible to exaggerate the importance of this step. Such a language even becomes a scientific instrument in itself (viz. its ability to predict the presence of hydrogen as a result of the reaction between zinc and hydrochloric acid). There was just one thing needed to underpin the new language – and this Lavoisier published two years later, in his *Elementary Treatise on Chemistry*. In this he defined the elements which would form the basis of this new chemical language, though at the same time he was aware of the insurmountable difficulties involved.

> I shall therefore content myself with saying that, if by the term elements, we mean to express the simple and indivisible molecules that compose bodies, it is probable that we know nothing about them: but if, on the contrary, we express by the terms elements or principles of bodies the idea of the last point reached by analysis, all substances that we have not yet been able to decompose by any means are elements to us; not that we can assert that these bodies that we consider as simple are not themselves composed of two or even a greater number of principles, but, since these principles are not separated, or rather since we have no means of separating them, they are to us as simple substances, and we must not suppose them compounded until experiment and observation have proved them to be so.

Essentially, this is no more than a refinement of Boyle's definition, made a century earlier. But there is a significant shift of attitude.

Here is the new pragmatism inspired by the improvements in, and reliance upon, experimental technique. Lavoisier is admitting that we might never discover precisely what an element is – we are forced to rely upon what appears to be an element, in the light of experimental practice. Chemistry was learning to admit what it did not know – and thus gain a deeper understanding of precisely what it did know. This too was a revolutionary advance in scientific method. Previously the elements had been defined theoretically. Earlier natural philosophers had been certain that they knew precisely what constituted an element, and what they were (earth, air, etc.). But this confident theoretical definition far outstripped any practical ability to confirm it. Now a definition was retained – but only as an ideal guideline, which also served as a healthy reminder of the inadequacies of our actual practical knowledge. It was not a straitjacket, like the concept of the four elements. And it even hinted at the possibility that such ideal theoretical definitions might ultimately be unattainable, and thus might have no place in an experimental science.

It is surely no coincidence that the greatest philosopher of the age was at this very moment formulating the metaphysical counterpart of such ideas. Almost a thousand miles away on the chilly shores of the Baltic in Königsberg, Immanuel Kant was outlining a philosophical world which consisted of two components: 'phenomena' and 'noumena'. The former were the appearance of things, as we perceive them through our senses, measurement and so forth. The latter were the unknowable core, the true world beyond the reach of our senses: the truth which supported and gave rise to these phenomena. Much of philosophy since has been an attempt to come to terms with this distinction – whose ramifications are both epistemological and scientific. Similarly, science too has remained a constant battle between theory-led and evidence-led progress – between attempts to grasp the inner truth and pinpoint the outer truth of the world around us. What exactly constitutes an element? Which are the elements?

Having proposed a provisional answer to the first question, Lavoisier now boldly listed his answer to the second. The list of the elements which Lavoisier drew up in his *Elementary Treatise on Chemistry* is surprisingly accurate. In all, he named thirty-three elements. Eight of these, including 'magnesia' and 'lime', have since been shown to be compounds (magnesium oxide and calcium oxide). Only two were utterly wrong. These were the elements he named as 'light' and 'caloric' (heat). Once thought to be material substances, these are now known to be forms of energy. Ironically, Lavoisier suggested that 'caloric' was an 'imponderable fluid' or principle, i.e. exactly the same as phlogiston. One mysterious 'principle', which also on occasion had 'negative weight', was replaced by another! Though its effect was not so fundamental as that of phlogiston, this idea of 'caloric' would hamstring research into heat for another half century.

Yet such cavils were far outweighed by Lavoisier's positive contribution. His definition of an element pointed the way forward for future exploration of the elements. Lavoisier may have discovered no new element himself (to his chagrin). Likewise, he was not solely responsible for any major discovery. No matter that he was not the sole destroyer of phlogiston, or the originator of the law of conservation of matter, or even the chemist who produced the definition of a chemical element (all of which he would have had us to believe). What he contributed was a revolutionary approach which established chemistry once and for all as a body of scientific knowledge about the real world.

But 1789 was also the date of another revolution. On 14 July the people of Paris stormed the Bastille. The French Revolution had begun – setting in motion the events which would eventually lead to the death of Louis XVI on the guillotine and the declaration of the republic. Lavoisier now found himself in an extremely tricky situation. As a scientist he had done as much as anyone to promote enlightened reform, but he was also a member of two institutions which were strongly linked with the *ancien régime*: the Académie des Sciences and the hated Ferme Générale. The initial euphoria of the revolution soon gave way to the inevitable blood-letting. Lavoisier remained Director of the Gunpowder Administration at the Arsenal under the new regime, steering his administrative career as best he could through the increasingly stormy and treacherous seas of revolutionary change. Ironically, it was his membership of the Académie des Sciences which brought him into danger.

Several years previously an ambitious young journalist had submitted a paper to the Académie, in the hope of gaining election to this prestigious body. The paper had been on the nature of fire. In it, the journalist claimed to have conducted an experiment which 'proved' how a burning candle in an enclosed space extinguished itself. According to the paper, this happened because the air heated by the flame expanded, and thus pressure mounted around the flame, diminishing its size, until finally it disappeared. Such an ingenious explanation might have merited some attention in the era of phlogiston, but in the light of Lavoisier's work it was of course sheer nonsense. It fell to the Academician Lavoisier to inform the misguided journalist that his paper was so devoid of scientific merit that, in modern parlance, it 'wasn't even wrong'. The journalist felt deeply insulted by Lavoisier's dismissive rejection; Lavoisier had made an enemy who would not forget him.

The journalist's name was Jean-Paul Marat. By 1791 Marat had become one of the leading members of the Jacobins, the extremist advocates of what would soon become the Terror. In 1791 Marat publicly attacked Lavoisier in the Jacobin newspaper *L'Ami du Peuple* ('The Friend of the People'). He described

Lavoisier as a 'charlatan . . . apprentice-chemist' who 'claims to be the legitimate father of every discovery. Since he has no ideas of his own, he supports himself entirely on the ideas of others.' This somewhat harsh view of Lavoisier's contribution may possibly have contained a grain of truth, but it soon became clear that Marat was interested in more than providing a corrective for posterity. He was much more interested in the present fate of this 'profiteer banker . . . chief tax collector . . . this little lord, who has an annual income of 40,000 livres'. Marat ended by declaring: 'Would that he had been strung from a lamp-post during the night.'

Within two years Marat led the Jacobins to power and the Terror was in full swing. Although Marat was soon assassinated, Lavoisier was arrested. Despite the frantic efforts of Mme Lavoisier, her husband was brought to trial. The judge expressed his opinion that 'The Republic has no need of scientists', and sentenced Lavoisier to death. He was guillotined the same day. When his fellow Academician, the celebrated French mathematician Joseph-Louis Lagrange, heard the news, his comment made a bitter epitaph: 'Only a minute to cut off that head, and a hundred years may not give us another like it.'

11

A Formula for Chemistry

Priestley heard the news of his great rival's death when he arrived in America. Yet despite Lavoisier's powerful experimental evidence, Priestley remained convinced to the end of his days that the phlogiston theory was right. As the twentieth-century German physicist Max Planck wryly observed: 'A new scientific theory does not triumph by convincing its opponents and making them see the light, but rather because its opponents eventually die.'

Meanwhile others were quick to build on the foundations of the new chemistry which Lavoisier had laid down. The most striking development came from the Englishman John Dalton. It has been claimed, with some justification, that Dalton contributed just one idea to science, and that the rest of his work was as mundane as the man himself. But the idea he contributed was the most profound and lasting ever incorporated into chemistry. It was not original, but Dalton's application of it was.

John Dalton was born in in 1766 in the remote village of Eaglesfield on the edge of the English Lake District – whose rugged beauty would be 'discovered' in a few years' time by the Romantic poet Wordsworth. All around lay a sublime beauty which had not yet been recognized. A fitting parallel: chemistry was in a similar state, and Dalton would be its Wordsworth. But Dalton was to be a poet of science, of precise intellect rather than precious rapture. His father was a hand-loom weaver of Quaker beliefs. Dalton himself left the local Quaker school at eleven, but returned a year later to teach there. His lessons were poorly delivered, and not surprisingly he had discipline problems with pupils older than himself. Few were perceptive enough to discern his enthusiasm, which, once stirred, became obsessive. It began with meteorology. The gaunt, gangling scholar became obsessed with recording the minute details of the daily weather: such were the prosaic daffodils of his inspiration. (Dalton would keep meticulous meteorological records for almost sixty years, taking his last readings on the very day he died. These were lovingly preserved for posterity until 1940, when the details of a wet Thursday afternoon in June 1796 – and countless other priceless observations – were blown to smithereens by a Nazi bomb.)

Dalton seems to have had a knack for wasting his scientific enthusiasm, and considerable talent, on inappropriate subjects. Despite the fact that he was colour-blind, his interest in meteorological phenomena led him to describe the aurora borealis. His poetic talent and defective eyesight reduced this awe-inspiring wonder to 'an elastic fluid, partaking of the properties of iron, or rather of magnetic steel, and that this fluid, doubtless from its magnetic property, assumes the form of cylindrical beams'.

Not until he was past thirty did Dalton seriously turn his attention to chemistry. By this time he was living a reclusive life in Manchester. Here he ran a small private tutorial college specializing in scientific subjects, where instruction was carried out with mainly home-made equipment. In the furtherance of his meteorological obsession he now began to study the composition of air, which in turn led him to a painstaking investigation of the behaviour and composition of gases. Accepting Boyle's notion that gases consist of tiny particles, he soon discovered a fundamental property of gases which to this day is known as Dalton's Law. This states that when two or more gases are mixed, their combined pressure will be the same as the added pressures of each gas if it was alone, occupying the same volume.

Just over a decade previously, in 1788, the French chemist Louis-Joseph Proust had discovered another important property of gases, which he found also applied to compounds of other substances. His law of definite proportions stated that all compounds consisted of elements in definite simple ratios by weight. In other words, a compound could contain two elements in the ratio 3:1, but not in such complex ratios as, say, 3.21:1 or 2.8:1. Dalton saw that this could easily be explained if Boyle's notion of gases was extended to all matter, so that all matter was viewed as consisting ultimately of tiny indivisible particles. If the particles of one element weighed three times that of another, and the compound was formed with one particle of one element joined to one particle of the other, then the ratio of their weights was bound to be precisely 3:1. It could never be 3.21:1 or 2.8:1.

Dalton recognized the similarity of these ultimate, indestructible particles to Democritus' idea of the 'uncuttable' *atomos*, and decided to call them atoms. But this was not simply a crib of the ancient Greek idea which had been so miraculously preserved and passed on through the works of Epicurus, then Lucretius, and finally through the sole surviving medieval copy of his *De Rerum Natura*. Nor indeed was it identical with any of the ensuing seventeenth- and eighteenth-century versions of this idea – such as Boyle's – none of which substantially advanced the Greek notion. All such ideas had remained utterly speculative and theoretical. Dalton's notion of the atom was scientific and practical. It was used to explain the experimental results which had led Proust to formulate his law of definite proportions. Democritus' concept had been theoretical. It had also posited the size and shape of atoms. (For instance, water atoms were smooth and round, causing water to be fluid and have no permanent shape.) Dalton's atoms, on the other hand, were concerned purely with weight. And although he had no way of determining the actual weight of atoms, he found he could determine their relative weights as they occurred in compounds. Dalton's was a quantitative theory which combined Democritus' original concept with Lavoisier's application of quantitative measurement to chemistry.

Dalton's atomic theory stated that all elements consisted of minuscule, indestructible atoms. Following on from Lavoisier, he held that all compound substances were simple combinations of these atoms. This momentous idea transformed our understanding of matter. During the two centuries following its discovery, science was to progress beyond all imagining. In the light of later discoveries Dalton's theory has been amended, but its basic premise remains fundamental to our present understanding of physics and chemistry. Indeed, the twentieth-century quantum physicist Richard Feynman asserted that if the human race was wiped out and could pass on just one sentence of scientific knowledge, this sentence should begin: 'All things are made of atoms . . .'

Having established that different elemental atoms had weights relative to one another, the next obvious step was to establish a marker. Hydrogen was the lightest element, so Dalton fixed this at a notional relative weight of 1. This meant that all other elements could be calculated relative to this figure. As an example, Dalton used water – which had already been established as a compound of oxygen and hydrogen, in proportions by weight of 8:1. Assuming that water consisted of one oxygen atom and one hydrogen atom, this meant that an oxgen atom weighed eight times as much as that of hydrogen. Dalton thus assigned oxygen an atomic weight of 8. (Here he was in fact mistaken: the atomic weight of oxygen is 16. Water contains two hydrogen atoms, but Dalton was not aware of this.) In this way he built up a table of atomic weights, listing each element with its weight in relation to that of hydrogen.

The supreme importance of Dalton's atomic theory was

quickly recognized throughout the scientific world. Yet apart from delivering a series of public lectures in his customary uninspiring style, Dalton continued to live a simple Quaker life in Manchester, shunning public honours. But these were showered upon him, whether he wanted them or not. He was elected to the Royal Society secretly, against his express wishes. Membership of such institutions was contrary to his religion. Notifications by post of his election to prestigious academies throughout Europe were not acknowledged.

World famous despite all his attempts to prevent this misfortune, Dalton finally died in 1844 at the age of seventy-seven. His wish for a simple Quaker funeral drew over forty thousand mourners and a hundred carriages. Britain had entered the Victorian age: respectability, reverence for famous worthies and pious ceremonies (especially funerals) were all central to the ethos of the rising middle class. Also, during the Industrial Revolution Manchester had risen from a small market town to become Britain's second largest city: the centre of its manufacturing industry, with a population of over a third of a million. (London, the world's largest city at the time, had a population of around two million. However, Manchester was the leading city in another sense: it was the first place on earth to mushroom into a massive sprawling urban area as a result of rapid industrial growth – a phenomenon which was to spread throughout the globe during the following century.)

Dalton's funeral was a celebration of civic pride, and also of the science which had made this possible. Science too had become respectable, even worthy. However, one of Dalton's wishes was respected. His eyes were preserved, in the hope that one day it would be discovered what had caused his colour-blindness. A hundred and fifty years after his death DNA samples showed that he had lacked the genes which produced green-sensitive pigment in normal eyes.

*

One further refinement was required before chemistry would be set free from the restraints of its own history. This was to be supplied by the greatest in the long line of Swedish chemists, Jöns Berzelius. While Dalton shunned honours, Berzelius seems to have enjoyed accumulating them. By the end of his illustrious career he had been honoured by no fewer than ninety-four academies, universities and learned societies, and the King of Sweden had made him a baron. By then his chemical textbook had been translated into all major languages and was regarded as the standard work, with his pronouncements on the latest chemical advances being taken as the holy writ. (Even when they were wrong, as they more frequently were in his conservative old age. In his opinion, neither chlorine nor nitrogen was an element, and that was that.)

But Berzelius' earlier work more than makes up for his inflexibility in old age. As a young man he was a surprisingly mediocre medical student, rescued from failure solely by the promise he showed at physics. Only towards the end of his studies did he begin to bloom as a chemist. But this breadth of scientific knowledge was to prove crucial. His first major work was done in electrochemistry, which he played a major role in developing. This new field had been made possible with the invention in 1800 of the electric battery, by the Italian Alessandro Volta, after whom the volt is named. Using the new 'voltaic pile' (as the battery was first called), Berzelius ran an electric current through solutions of different compounds. This caused them to separate, with one part attracted to the anode (positive terminal) and one to the cathode (negative). For instance, with copper sulphate the copper would be attracted to the cathode. As negative electric charge attracts positive, this made Berzelius realize that the copper component of copper sulphate had a positive charge. This process came to be known as electrolysis (literally 'electrical-unbinding'). Further experiments on other compounds produced similar results, prompting Berzelius to propose a far-reaching theory. It appeared that all compounds were dualistic, consisting of a positive and a negative component, held together by their opposing electrical charge.

In this way Berzelius was able to make up a list of elements – with oxygen, the most negative, at one end, and highly positive alkali metals at the other. He had discovered an entirely new way of listing the elements, one which appeared to bear no precise relation to their atomic weights. Berzelius had been one of the earliest to accept Dalton's atomic theory, and this had led him to embark upon an exhaustive exploration of atomic weights. By 1810 Dalton had managed to establish the atomic weight of twenty elements. Berzelius found that Dalton's figures were of varying accuracy. (Colour-blindness, the dogged insistence upon constructing his own apparatus and native clumsiness limited Dalton's effectiveness as an experimenter: his forte was his ability to perceive a theoretical pattern in a mass of data.) Berzelius, on the other hand, was a persistent and meticulous experimenter. By 1818 he had determined atomic weights for forty-five of the forty-nine accepted elements. At the same time, he had also analysed over two thousand compounds in the attempt to confirm his dualistic theory. Unfortunately he had found that certain compounds did not appear to possess this dualistic positive/negative nature. This was particularly the case with organic compounds – those which contain carbon, often in complex structures, and form the basis of living organisms.

However, this anomaly did not cause Berzelius to abandon his dualistic theory, which still appeared to provide the key to chemical reaction. Instead he insisted that organic compounds, because they were living, were subject to a 'life-force' which functioned over and above the laws of chemistry. This doctrine, known as vitalism, was to remain remarkably persistent. The similarity between the 'life-force' and phlogiston is evident. Its inability to show up in any experiment left it vulnerable to the opposing theory of materialism, which states that everything that exists either is, or depends upon, matter. Nowadays the 'life-force' may have been eliminated from science, along with other manifestations of the spirit world, but materialism too is not without its difficulties. Everything may consist of matter, but what exactly is this 'matter'? And how can we ever know anything for certain about it? How can our sense organs, and their extension in the form of scientific instruments, possibly give us direct access to matter? We see with our eyes, not through them. The instrument registers only what it is constructed to register, which does not necessarily resemble the actuality with which it is dealing. Indeed, what it registers certainly cannot be identical with that actuality. Which returns us to the problem foreseen by Lavoisier, when he defined an element – the philosophical problem raised by Kant with his phenomena and noumena. We don't know for certain. When science uses explanations or theories which go beyond experimental evidence, it leaves itself open to question. Science is what works, not a philosophical explanation of the world.

But does this mean science must always rely upon experimental results, forgoing such theories as phlogiston or even 'life-force', which may, for a limited period, provide fruitful explanations? During Berzelius' time chemistry began to rely upon one particular entity, which became an essential component of all experiments, yet was never supported by any experimental evidence whatsoever. And this lasted almost as long as phlogiston and the 'life-force'. The entity concerned was the atom. For a century after Dalton proposed his atomic theory, no one was able to provide concrete evidence that such a thing as an atom actually existed. As late as the early years of the twentieth century, the non-existence of the atom was still being plausibly maintained by such eminent thinkers as the Austrian philosopher-scientist

Ernst Mach (after whom Mach 1, the speed of sound, is named). These were no flat-earthers. They recognized that the concept of the atom had been of immense value to science through the preceding century – but insisted, quite correctly, that it remained just a concept. Science may be 'what works', but the foundations upon which it is based still remain in part somewhat flimsy. (Not until Einstein's paper in 1905 was the existence of the atom actually proved.) Alchemy achieved a great deal for chemistry during a time when science was progressing at a tortoise pace, though its wizardry is now seen as laughable. With contemporary science streaking ahead like a hare, our present unrealized alchemies and unjustified atom-like assumptions will doubtless be exposed to ridicule much sooner. We will all appear flat-earthers to our grandchildren.

Berzelius' painstaking analysis of chemical compounds eventually led him to discover three new elements (serium, silenium and thorium). His faithful team of assistants added half a dozen more. But not all great advances in science come through discoveries, or even original conceptualizations (supported or otherwise). Lavoisier had set out the infrastructure of chemistry. Berzelius added the finishing touches to this project. Chemistry was now well established as an international science – yet unlike mathematics, for example, it had no international language. When Lavoisier decreed that compounds should be named after their elemental constituents, this project was hampered by the fact that elements often had well-established different names in different countries. For instance, in Germany hydrogen was (and still is) called *Wasserstoff* ('water-stuff', a German version of Lavoisier's Greek for 'water-generator'). Lavoisier had indicated the way forward by forming names for new elements, such as oxygen and hydrogen, from ancient Greek descriptions of their distinctive properties. Berzelius used his authority to promulgate this notion throughout the scientific world, insisting that in scientific papers the elements should be called by their ancient Greek or Latin names. Thus gold (*or* in French, *guld* in Swedish) became the Latin *aurum*; and silver (*argent* in French, *Silber* in German) became the Latin *argentum*.

But this was only the first step. From earliest times alchemists had represented chemical reactions by formulae, using secret symbols, hieroglyphs and pictographs to depict the starting ingredients and the end products. Lavoisier had understood the usefulness of such formulae, as long as the symbols used were known to all. Unfortunately, the symbols he adopted were almost as impenetrable as the alchemists' hieroglyphs. Dalton understood the need for a much simpler symbolism. Since he visualized atoms as tiny circular entities, he understandably chose to represent the elements in circular form. Hydrogen was represented by a circle with a dot in the middle; sulphur had a cross in the circle; mercury had dots around the inner circumference, making it look like a cogwheel; copper had a c in the circle, like a copyright sign. Compounds were shown as groups of inscribed circles joined together, in appropriately attached clusters. This produced complex patterns of striped, spotted, dotted and shaded circles – resembling anything from a Michelin man to a formation of snooker balls. These patterns had the merit of a certain pictorial accuracy – but would have baffled many an expert cryptographer, let alone a chemist trying to read off a chemical equation.

It was Berzelius who saw the simple answer. He decided that in all chemical equations the element should be represented by the initial letter of its classical Greek or Latin name. For example, hydrogen should be H, oxygen O, and so forth. Where two elements had the same initial, a second distinguishing letter from the classical name should be added. Thus aurum (gold) became Au, and argentum (silver) became Ag. Compounds could now be written, rather than depicted, in simple symbolic form. For instance, carbon monoxide could be written CO. And when more than one atom was found to be present in a compound, it was decided that this should be indicated by a subscript number. Thus carbon dioxide should be written CO_2; and ammonia (which contains one nitrogen atom and three hydrogen atoms) became NH_3.

Chemistry at last had its own universal language, like mathematics. And this was, in its own way, mathematical. Unlike Lavoisier's descriptive nomenclature, which could predict only what chemicals would result from a reaction, this new mathematical formulation could also predict the relative quantities that would be produced. For instance, Lavoisier's descriptive nomenclature showed that:

zinc + hydrochloric acid = zinc chloride + hydrogen.

But Berzelius' formula showed the precise relative proportions required for (and produced by) this reaction:

$$Zn + 2HCl = ZnCl_2 + H_2$$

Chemical formulae, just like mathematical formulae, had to balance out.

For chemistry, this was the equivalent of mathematics changing from Roman to Arabic numerals (when the opacity of XL × V = CC gave way to the clarity of 40 × 5 = 200).

Mathematics had now entered the very heart of chemistry, enabling it to see precisely what it was doing.

12

The Search for a Hidden Structure

In the wake of Lavoisier, the systematic approach and new experimental techniques soon led to the discovery of a host of new elements. During the life of Berzelius (1779–1848) no fewer than thirty-two new elements were isolated, bringing the total up to fifty-seven. The British scientist Sir Humphry Davy alone discovered six elements. The most important of these were isolated by electrolysis. In October 1807 Davy built the most powerful battery so far assembled, using 250 plates. This enabled him to pass a strong electric current through an aqueous solution of potash, a compound which he had long suspected of containing an unknown element. At first the current only caused the water to decompose, so instead he eliminated the water and repeated the experiment with an igneous fusion of potash (i.e. molten). In this way Davy managed to separate tiny globules of an alkali metal, which he named potassium. When a globule of it was introduced to water it burst into flame and scudded about the surface issuing a fierce hissing sound. In chemical terms, this showed that the isolated form of this metal was so reactive that it extracted oxygen from the water, liberating a hiss of hydrogen gas which burst into flame due to the heat of the reaction. Later in the same week Davy isolated another alkali metal by electrolysis, this time from caustic soda. This he called sodium. At the time, the discovery of these highly reactive alkali metals caused almost as much of a sensation as the discovery of phosphorus a century and a half previously – and for similar reasons.

Scientific lectures were once again all the rage amongst polite society, and a spectacular 'demonstration' of one these newly discovered elements always made a stir, frequently causing a few of the ladies in the audience to faint.

The enforced idleness of intelligent but improperly educated middle-class women, and their consequent hunger for knowledge, meant that these events provided something of a popular education, and were always well attended by the ladies. Yet any deeper participation in science was still considered socially unacceptable for women. However, Mme Lavoisier's pioneering example was to be followed by a few intrepid souls. Notably: Ada Lovelace, Byron's neglected daughter, who wrote the first programme for Babbage's original 'analytical engine' computer; Sophie Germain, who, despite being largely self-taught, attracted the attention of the great Gauss and remains the finest woman mathematician France has yet produced; and Caroline Herschel, the German-British astronomer who discovered no fewer than eight new comets and revised John Flamsteed's classic *Observations of the Fixed Stars* for the Royal Society (though she was of course not allowed to become a member). All this only serves as an indication of what might have been, if the vision of science had not moved forward with one eye firmly closed.

Aluminium is now known to be the most frequently occurring metal in the earth's crust – but for centuries it remained undetected. The brilliant German chemical theorist Georg Stahl, who developed and named phlogiston theory, was probably the first to suspect that there was a hitherto unknown element in alum (aluminium potassium sulphate). However, it was to be a century and a half before his hunch was confirmed. In 1827 the German chemist Friedrich Wöhler finally succeeded with supreme experimental ingenuity in isolating metallic aluminium. (Basically, Wöhler's experiment involved heating dehydrated aluminium chloride with hyper-reactive pure metallic potassium, which stripped the chlorine from the aluminium.) Wöhler then set about examining the properties of this silvery-white crystalline metal.

Wöhler's method of isolating aluminium was to prove so difficult, and the properties of this new metal so glittering, that for a few decades aluminium became more valuable than gold. Thirty years after its discovery, when a shining bar of aluminium was exhibited in Paris, Napoleon III ordered a set of cutlery made of this new metal. His intention was to entertain the crowned heads of Europe with knives and forks which nowadays wouldn't even be acceptable in a prison canteen.

Besides isolating aluminium, Wöhler's other great achievement was his synthesis of urea, which is a product of living matter, from non-living materials. This creation of organic matter from inorganic matter disproved the widely held theory of vitalism (and the 'life-force'), though it would be many obstinate years before it was accepted that life did not exist.

Several elements were being discovered almost every decade. This profusion of new elements with an ever-widening range of properties soon began to provoke questions. Precisely how many elements were there? Had most of them already been discovered? Or would there perhaps turn out to be innumerable elements? This soon led to more profound speculations. Somehow, amongst all these elements, there must be some kind of fundamental order. Dalton had discovered that the atoms of each element had different weights – but surely there had to be more to it than this? Berzelius had noticed that elements appeared to have different electrical affinities. Likewise, there appeared to be groups of different kinds of elements with similar properties – metals which resisted corrosion (such as gold, silver and platinum), combustible alkali metals (such as potassium and sodium), colourless, odourless gases (such as hydrogen and oxgygen) and so forth. Was it possible that there was some kind of fundamental pattern behind all this?

Chemistry had achieved its scientific status and continuing success largely through experiment, and such theoretical thinking was viewed at best as mere speculation. Why should there be some kind of order amongst the elements? After all, there was no real evidence for such a thing? But the rage for order is a basic human trait, not least amongst scientists. And these speculations eventually began to find support, if only from scraps of evidence.

The first of these came from Johan Döbereiner, the professor of chemistry at the University of Jena. Döbereiner was the son of a coachman, and was largely self-educated. He managed to obtain a post as a pharmacist, and eagerly attended the regular local public lectures on science. His precocious chemical knowledge brought him to the attention of Karl August, Duke of Weimar, who eventually secured his appointment at the University of Jena. Here his lectures were regularly attended by Goethe, who had such a compelling interest in science that at one stage he considered his activities in this field more important than his writing.

It is worth examining Goethe's interest here, as it is indicative of a level and range of amateur scientific interest which was not uncommon amongst educated people of this period, both throughout Europe and in America. Goethe's scientific speculations are better known because of his literary genius, and for this reason his wrong-headedness in this department is often accorded a respect it ill deserves. Despite Newton having amply demonstrated that white light contained light of all colours, Goethe insisted on maintaining that it was a colour in its own right. He held that all colours were in fact a mixture of light and darkness, infused with a cloudy medium which lent the ensuing grey dusk its coloured radiance. (Later the philosopher Schopenhauer, a scientific thinker of some merit who should have known better, would also champion this fantasy.) Goethe's other scientific adventures included a prolonged search for the 'ur-plant' from which all others had developed, as well as the invention of 'morphology', the study of the 'unity' underlying the diversity of all animal and plant life. Such notions were purely speculative, based on little more than the insights of a fertile imagination. (Even so, the resemblance to the belief that there was a pattern behind the elements is remarkably similar.) Goethe was wrong, and is thus open to the ridicule of hindsight. Yet it is not difficult to see in his thinking a tentative step towards the idea of evolution, which Darwin would formulate less than a quarter of a century after his death.

Goethe was not alone in taking his hobby seriously. Amongst thinking men (and a few pioneer women) theoretical scientific speculation was becoming increasingly widespread, encouraged more than a little by the achievements of the Industrial Revolution. But this same revolution also brought its dark satanic mills and the prospect of a future filled with squalor and social disruption. Likewise the scientifically minded amateur working alone in his laboratory at the outreaches of theory also acquired his dark side, which played on the fears of the unknown. Even while Goethe was in his prime, Mary Shelley was writing *Frankenstein* – an ur-figure of the mad scientist and his demonic creation whose potency persists to this day.

Meanwhile Goethe's chemistry teacher, Professor Döbereiner, was working on his own morphological ideas. In 1829 he noticed that the recently discovered element bromine had properties which seemed to lie midway between those of chlorine and iodine. Not only that, its atomic weight lay exactly halfway between those of these two elements.

Döbereiner began studying the list of the known elements, recorded with their properties and atomic weights, and eventually discovered another two groups of elements with the same pattern.

Strontium lay halfway (in atomic weight, colour, properties and reactivity) between calcium and barium; and selenium could be similarly placed between sulphur and tellurium. Döbereiner named these groups triads, and began an extensive search of the elements for further examples, but could find no more. Döbereiner's 'law of triads' appeared to apply only to nine of the fifty-four known elements, and was dismissed by his contemporaries as mere coincidence.

And that was it, for the time being. Chemistry had suffered enough from mistaken theories (four elements, phlogiston, etc.). The way forward now lay through experiment.

It would be over thirty years after Döbereiner's law of triads before another significant attempt was made to discover a pattern in the elements. Unfortunately, this contribution was to come from a scientist whose brilliance was matched only by his waywardness. Alexandre-Emile Béguyer de Chancourtois was born in Paris in 1820. His first love was geology, which took him on expeditions as far afield as Turkestan, Armenia and Greenland. He returned convinced that it was the geology of a country which determined the lifestyle of its inhabitants. In other words, it was its deposits of coal or sulphur, rather than, say, the climate, social structure or racial characteristics, which were the major influence on local behaviour. Yet this unpromising start proved to be the beginning of human geography, and de Chancourtois is now recognized as one the founders of the subject. Later, he was appointed inspector general of mines in France, where his unorthodox approach led him to introduce sweeping safety measures and modern engineering methods at the expense of the outraged mine-owners. De Chancourtois didn't turn his considerable talents to chemistry until he was in his forties. In 1862 he produced a paper describing his ingenious 'telluric screw', which demonstrated that there did indeed appear to be some kind of pattern amongst the elements. De Chancourtois' 'telluric screw' consisted of a cylinder on which was drawn a descending spiral line. At regular intervals along this line de Chancourtois plotted each of the elements according to its atomic weight. He was intrigued to find that the properties of these elements tended to repeat when the elements were read off in vertical columns down the cylinder. It seemed that after every sixteen units of atomic weight the properties of the matching elements tended to exhibit striking similarities with those vertically above them on the cylinder. De Chancourtois' paper was duly published, but unfortunately he chose to revert to geological terms when referring to certain elements, and at one stage even introduced his own version of numerology (the alchemy of mathematics, in which certain numbers have their own esoteric significance). To make matters even worse, the publishers omitted to include de Chancourtois' illustration of the cylinder, thus rendering the article virtually incomprehensible to all but the most persistent and informed reader. (As we shall see, there was to be just one person who fell into this category. But this reader would be so inspired by de Chancourtois' work that he would go on to transform the face of chemistry.)

This subject evidently attracted a certain type of scientific thinker inured to ridicule. In 1864 the young English chemist John Newlands came up with his own pattern of the elements, unaware of de Chancourtois' cryptic researches. John Newlands was born in London in 1837, the son of a Presbyterian minister. His mother was of Italian descent, a fact of which he was particularly proud. At the age of twenty-three Newlands interrupted his scientific studies to sail for Palermo, where Garibaldi had raised the Italian flag to herald the liberation and unification of Italy. Here Newlands volunteered for Garibaldi's army, the celebrated Red Shirts. Italy would be united and ruled by Italians for the first time since the Roman Empire. Europe was now solidifying into large national power blocs: the unification of Italy was concurrent with the unification of Germany, which had been fragmented since the Reformation. Modern Western Europe was beginning to take shape in the wake of the Industrial Revolution, with steel-processing plants, mining industries and chemical factories being established from Sweden to Greece.

On his return from Italy Newlands began a study of the elements. He came up with findings that bore a certain resemblance to those of de Chancourtois, though they were a significant advance on the Frenchman's ideas. Newlands discovered that if he listed the elements in ascending order of their atomic weights, in vertical lines of seven, the properties of the elements along the corresponding horizontal lines were remarkably similar. As he put it: 'In other words, the eighth element starting from a given one is a kind of repetition of the first, like the eighth note in an octave of music.' He named this his 'law of octaves'. In the tabulated list the alkali metal sodium (the 6th heaviest element) stood horizontally beside the very similar potassium (13th heaviest). Likewise, magnesium (10th) was in line beside the similar calcium (17th). When Newlands expanded his table to include all the known elements he found that the halogens, chlorine (15th), bromine (29th) and Iodine (42nd), which exhibited graduating similar properties, all fell in the same horizontal column. Whereas the trio of magnesium (10th), silenium (12th) and sulphur (14th), which also had graduating similar properties, fell in the same vertical line. In other words, his law of octaves also seemed to incorporate the scattered resemblances noted in Döbereiner's law of triads. Unfortunately Newlands' tabulated law of octaves also had its faults. The properties of some elements, especially those of higher atomic weight, simply didn't tally. Even so, Newlands' law of octaves was a definite advance on any previous ideas. Indeed, many now regard it as the first solid evidence that there was indeed some comprehensive pattern to the elements. In 1865 Newlands reported his findings to the Chemical Society in London, but his ideas proved ahead of their time. The assembled worthies merely ridiculed his law of octaves. Amidst the general merriment, one even asked him sarcastically if he had tried arranging the elements in alphabetical order. It would be a quarter of a century before Newlands' achievement was finally recognized, when the Royal Society awarded him the Davy Medal in 1887.

Döbereiner had spotted resemblances between isolated groups of elements. De Chancourtois had discerned a certain pattern of recurrent properties. Newlands had extended this pattern and even incorporated Döbereiner's groups. But still his law of octaves didn't work overall. This was partly due to contemporary miscalculations of various atomic weights and partly because Newlands made no allowances for hitherto undiscovered elements. But it was also because the rigidity of Newlands' octave system just didn't fit.

It was becoming increasingly obvious that there was some kind of pattern to the elements, but the answer was evidently more complex. Chemistry appeared to be tantalizingly close to glimpsing the blueprint of the very elements upon which it was based. Euclid had laid the foundations of geometry, Newton's gravity had explained the world in terms of physics and Darwin had accounted for the evolution of all species – could chemistry now discover the secret which accounted for the diversity of matter? Here, possibly, was the linchpin which could unite all scientific knowledge. The next man who attempted to solve this problem was the possessor of the finest chemical mind since Lavoisier.

13

Mendeleyev

Dmitri Ivanovich Mendeleyev was born on 8 February 1834. Or 27 January, according to the old Julian calendar, which was still in use in Russia. By now the rest of Europe had adopted the Gregorian calendar, leaving Russia twelve days behind the rest of the world. This backwardness was symptomatic of an entire culture. In the 1830s Russia existed for the most part in feudal isolation, the majority of its inhabitants serfs who remained the unpaid property of the landowners. The tsar (a name which derived from Caesar, like the Julian calendar) was God's representative on earth, and ruled by divine right. There had been no Reformation (or even a renaissance) in imperialist Russia.

Dmitri Ivanovich Mendeleyev was born in Tobolsk in western Siberia, the youngest of fourteen or seventeen children (no one seems to know which). His father was headmaster of the local gymnasium but became blind in the year of Dmitri's birth, leaving the mother to provide for the large family. Fortunately, Maria Dmitrievna, née Kornilov, was an exceptional woman. The Kornilov family were merchants who had played a leading role in opening up western Siberia. Her father had established paper and glass factories at Tobolsk. Less than fifty years previously he had opened the first printing press in Siberia, and had launched the first newspaper in the four-thousand-mile-wide province. One of Maria's ancestors had married a local Khirgis Tartar beauty – and given such ample scope for genetic expression it was not surprising that a few of Mendeleyev's siblings exhibited Mongolian features, though he himself did not.

To provide an income for the family Maria reopened her father's glass factory, which was in a remote village twenty miles north of Tobolsk. Here she built a wooden church for the workers and set up a school to educate their children. Mendeleyev's earliest memories were of the huge red glow from the glass furnaces lighting up the night sky above the endless darkness of the Siberian woods.

Mendeleyev went to school in Tobolsk, where he did badly. In those days, education consisted of a thorough grounding in dead languages. Ancient Greek and Latin were meant to give an understanding of the classical ideals upon which civilization was based. Such notions were about as relevant in mid-nineteenth-century Siberia as Shakespeare and Goethe seem to many in modern life, and Mendeleyev developed a distaste for high culture which was to last a lifetime. Fortunately he received some private tutoring from an exiled Decembrist called Bessagrin, who had married an older sister.

The Decembrists were the remnants of the unsuccessful 1825 December Revolution, which had been staged by a group of liberal intellectual army officers (curiously, not a contradiction in terms in this instance). The revolution, with its inflammatory demands for constitutional government, had quickly been put down. Afterwards the ringleaders had been sentenced to death, and Bessagrin and the rest of his colleagues banished to Siberia.

Bessagrin instilled in Mendeleyev a deep interest in science, as well as reinforcing the liberal ideals put into practice by his mother. It soon became clear that Mendeleyev had an exceptional mind, and he began pursuing his own schoolboy experiments. (In later life Mendeleyev liked to play up his Siberian origins, claiming that he had been brought up amongst primitive Tartars in far eastern Siberia, and had spoken no Russian until he was seventeen. Owing to his outlandish hirsute appearance, this story was frequently accepted without question.)

In 1847 the family was struck by a sequence of catastrophes. Mendeleyev's father died, and the following year the glass factory burned down. In 1849, when Dmitri Mendeleyev was fifteen, his mother set off with her two remaining dependent children – Dmitri and Liza – for Moscow. This meant a laborious 1,300 mile journey, often involving rides hitched on horse-drawn wagons. Maria Mendeleyeva was now fifty-seven years old, tired and aged well beyond her years after bringing up her huge family single-handed, while at the same time running a factory and organizing the welfare of its workers. But she was determined that her brilliant Dmitri should receive the education his promise deserved.

In Moscow Mendeleyev's application to enter the university fell foul of the bureaucracy. Entry from the provinces was according to a quota system; but the pioneer province of Siberia had not yet been given a quota. Mendeleyev was refused entry. When his mother applied to other institutions of further education, she was informed that his Siberian qualifications were simply not recognized in Moscow. As a last resort, the Mendeleyevs set off a further four hundred miles for the capital, St Petersburg.

Here it was much the same: catch-22 regulations administered by Kafkaesque officials. Fortunately Maria discovered that the head of the Central Pedagogical Institute, the main training college for high-school teachers, was an old friend of her dead husband. Where bureaucracy disappointed, favouritism appointed. Mendeleyev was given a place to study mathematics and natural science, as well as a small government scholarship sufficient to provide him with necessary support.

Within ten weeks of Mendeleyev's entering the Central Pedagogical Institute, his mother was on her deathbed. Her last words to her favourite son were typically forceful: 'Refrain from illusions, insist on work and not on words. Patiently seek divine and scientific truth.' Mendeleyev never forgot these words. Thirty-seven years later he would quote them in a scientific treatise which he dedicated to her memory, adding: 'Dmitri Mendeleyev regards as sacred a mother's dying words.'

Just over a year after the death of Mendeleyev's mother, his sister Liza also died. A year later Mendeleyev suffered a throat haemorrhage and was placed in the institute hospital. He had never been a healthy child, and tuberculosis was now diagnosed. The doctors reckoned he had only a few months to live.

Having reached this Dickensian nadir, Mendeleyev proceeded to play the sentimental role required. He spent long periods in bed, and appears to have been regarded as a sort of mascot by the institute. The orphan was adopted by science. Mendeleyev would rise from his bed to work in the institute laboratories, and was soon performing original experiments. He would retire to bed to write up his work in papers which he sent to St Petersburg's scientific journals. Some examples of this original work were published while he was barely twenty and still an undergraduate. The originality of these papers was due to Mendeleyev's talent, yet this could not have expressed itself fully but for the exceptional teaching at the Central Pedagogic Institute. This was housed in the same buildings as St Petersburg University, and many of the university professors also doubled as lecturers in the institute.

The city of St Petersburg had been created out of nothing amidst the marshes of the Baltic shoreline some 150 years previously by Peter the Great. He had intended it as Russia's 'window to Europe'. By the 1850s St Petersburg was beginning to emerge as one of the intellectual centres of Europe, and St Petersburg University was the finest academic centre in the land, with a strong science faculty. The professor of physics was Emil Lenz, now best rememberd for Lenz's Law, one of the fundamental principles of electromagnetic induction. The chemistry department was headed by A. A. Woskressensky, whose lectures on the chemical elements were renowned for their comprehensiveness. Now that new elements were being discovered every few years, chemistry was beginning to displace physics as the science which caught the public imagination. Woskressensky's lectures even extended to detailing the properties of such rare elements as uranium and ruthenium (both of which had been discovered in the previous decade). Mendeleyev was enthusiastically absorbing an encyclopedic wealth of chemical lore, but his mind did not become bogged down in all these facts. Even in these early days he excelled in the ability to link apparently isolated pieces of knowledge. It was this seemingly whimsical knack which accounted for the originality of his scientific papers. But beneath this superficial brilliance was developing a far more profound talent: the ability to see a pattern in a wealth of apparently unrelated material.

In 1855 Mendeleyev qualified as a teacher, taking the gold medal for the best student of his year. His first appointment was to a teaching post at Simferopol in the Crimea. This is usually put down to the benevolence of the authorities: the temperate southern climate would be good for his health. In fact, the opposite was the case. Mendeleyev may have liked to play the institute mascot, but he also had a less lovable side. When thwarted he had a short fuse. And in his tantrums he was capable of working himself up to the point of dancing with Rumpelstiltskin-like rage. On one occasion he apparently vented this rage on an official of the Ministry of Education. Revenge is a dish best enjoyed cold: the official bided his time, and when Mendeleyev graduated he had him posted to Simferopol. Mendeleyev set off for the sun with high hopes.

He arrived at Simferopol to find the Crimean War in full swing. The entire region had been transformed into a vast army camp, and the gymnasium at Simferopol had been closed for months. Mendeleyev found himself stranded, without a job, with no prospect of being paid. The ensuing Rumpelstiltskin dance beneath the blazing sun can't have been good for a man in his condition.

However, good was to come of all this. It was in Simferopol that Mendeleyev met the man he was to regard as his saviour. The renowned surgeon Perogov was working at the local military hospital. He gave Mendeleyev a check-up and diagnosed his disease as non-fatal. Fired with enthusiasm at this news, Mendeleyev soon returned to St Petersburg. Here, at the exceptional age of twenty-two, he was appointed as a *privat Dozent* (untenured, unpaid lecturer, who depended upon fees from students attending his course) in the University of St Petersburg. Meanwhile Mendeleyev continued with his researches in the university laboratories, but became increasingly frustrated.

For all its new-found cultural status, St Petersburg in fact remained backward in many fields. There was still practically no opportunity for advanced scientific research here, or anywhere in Russia. In 1859 Mendeleyev managed to secure a government grant to study abroad for two years. On the advice of his friend, the chemist and composer Borodin, Mendeleyev headed first to Paris, where he studied under Henri Regnault, the finest experimentalist of his era. It was Regnault who first established that absolute zero was – 273°C, and he almost certainly achieved many other experimental results far ahead of his time. Unfortunately, his extensive laboratory notebooks were to be destroyed during the anarchy when the Communards took over Paris in 1871. Regnault was unable to reconstruct this experimental work before his death eight years later, but maintained to the end that he had discovered the principle of conservation of energy well before Joule.

After Paris, Mendeleyev set off for Heidelberg. Here he briefly attended lectures by Gustav Kirchhoff, said to have been the most boring lecturer in all Germany at the time. (Some feat, considering this was the high era of German metaphysics, when prolix philosophers prided themselves on delivering lectures replete with page-long sentences.) But away from the sound of his own voice, Kirchhoff was a chemist of prime brilliance who was to discover a number of new elements.

After giving up on Kirchhoff's lectures, Mendeleyev worked briefly at Heidelberg with Kirchhoff's great partner Robert Bunsen, best remembered today as the inventor of the Bunsen burner, which is still found in every school lab. But Bunsen's major work was the development of spectroscopy with Kirchhoff. This was to prove a major tool in the identification of new elements.

Between them, Kirchhoff and Bunsen developed the spectroscope, which uses a prism to refract light. As Newton had shown, when light passes through a prism the different wavelengths of which the light consists are refracted in varying degrees, so that it breaks up into a spectrum of its constituent colours. White light, for instance, becomes a rainbow:

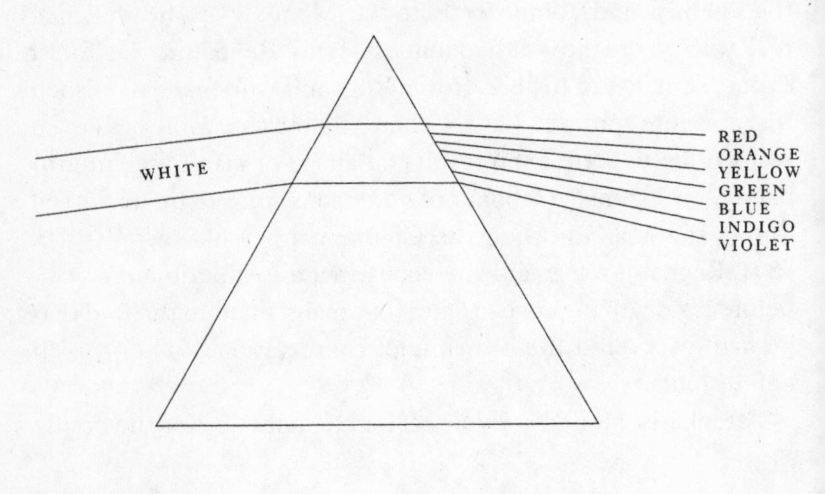

Kirchhoff and Bunsen found that when an element was heated, the light it emitted produced its own characteristic spectrum of coloured lines. Each element had its own 'fingerprint', so to speak.

This research was much assisted by Bunsen's new burner, whose flame produced a minimum of background light. This ensured that the 'fingerprint' spectrum of the element being heated in its flame was not 'blurred' by other light. In 1859, a matter of months before Mendeleyev's arrival in Heidelberg, Kirchhoff and Bunsen heated a compound which produced some deep red spectral lines matching those of no known element. They had discovered the presence of a new element, which they named rubidium (the Latin for 'deep red').

Kirchhoff now extended this method by a brilliant insight. When light passed through a gas, certain dark lines appeared in its spectrum. Kirchhoff found that these dark lines precisely matched those in the spectrum which the gas would have produced when heated. It appeared that the gas absorbed the spectral lines of its own fingerprint, producing a sort of negative spectrum consisting of dark bands.

When Kirchhoff had studied sunlight with his spectroscope, he had detected a number of unaccountable dark bands in its spectrum. The light from the sun had to pass through its atmosphere, and he realized that these dark bands were the 'fingerprints' of the elements in the sun's gaseous atmosphere. Using this method, he discovered that the sun's atmosphere contained sodium vapour. This meant that one of the sun's constituents was sodium. Continuing with this spectroscopic method, Kirchhoff went on to discover half a dozen new elements in the sun which had not so far been found on earth. Science was beginning to take previously unimaginable steps, far beyond the realms of the known world. (Just a quarter of a century earlier, the French positivist philosopher Auguste Comte had pronounced that certain kinds of knowledge would remain for ever beyond the reach of science. For instance, it would never be possible to discover precisely what the stars were made of. Comte died mentally unhinged in Paris, just a few years before Kirchhoff unhinged his philosophy.)

Mendeleyev found himself in the right place at the right time. At Heidelberg his knowledge of the chemical elements benefited greatly from the discoveries going on around him. Working with Bunsen, he had privileged access to the very latest developments.

But here once again Mendeleyev's temperament proved his undoing. He didn't hit it off with Bunsen. After another tantrum, Mendeleyev stormed out of the Heidelberg laboratories, vowing not to return. The fact that he had effectively disbarred himself from the finest chemical laboratories in Germany, and nullified the entire purpose of his visit to Europe, does not seem to have bothered him. Instead, he transformed one of the two rooms in his lodgings into a makeshift private laboratory, and continued his researches at home. Here he was limited to experiments concerning the seemingly mundane problem of the solubility of alcohol in water.

But Mendeleyev's ability to detect resemblances between apparently disparate findings now underwent a significant development. He evolved an uncanny ability to detect an underlying principle amidst a welter of seemingly commonplace findings. What began as a study of the solubility of alcohol in water (the subject of his doctoral thesis) soon developed into an investigation of the nature of solutions – which led him to a deeper study of atoms, molecules and valency.

Some elements, such as hydrogen, do not exist in independent atomic form (H). Hydrogen gas, for instance, consists of atoms combined in pairs (H_2). Such combinations of atoms are known as molecules – which can be made up of similar or dissimilar elements. For instance, two hydrogen atoms (H) combine with one oxygen atom (O) to form a molecule of water (H_2O). The valency of an atom is the measure of its ability to combine with other atoms. Imagine an atom as a ball, then its valency can be pictured as the amount of protruding arms it has which enable it to link up with other atoms. For example, hydrogen has a valency of 1, and oxygen a valency of 2. So:

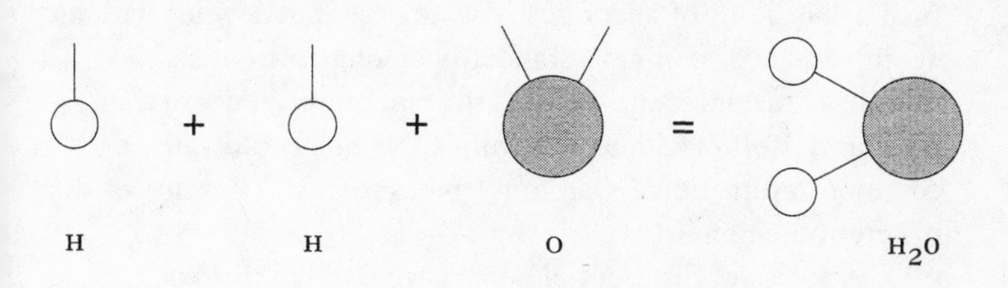

The valency of an element is as fundamental as its properties. It too is a defining characteristic of an element. Mendeleyev grasped this vital point early in his seemingly minor study of the solubility of alcohol in water.

This ever-deepening study also led him to an important discovery about gases and their properties. When temperature is decreased and pressure increased, a gas will liquefy. (This is how carbon dioxide is dissolved in water to make fizzy drinks. We have all witnessed the opposite of this process: every bottle of Coke contains liquid carbon dioxide, which fizzes into gas once the cap is removed and the pressure decreases. And when the temperature is higher, the Coke fizzes even more.) Working alone in his laboratory in Heidelberg, Mendeleyev discovered that every gas has a critical temperature. If the gas is heated above this temperature, no amount of pressure can turn it into liquid.

This discovery is usually credited to the Irish chemist Thomas Andrews, who in fact discovered critical temperature two years later. Working in obstinate isolation in his private laboratory, Mendeleyev's findings did not attract sufficient attention to ensure his priority. Not even Andrews noticed Mendeleyev's report – there is no question of plagiarism here.

Science continues to develop regardless of disputes over priority. But disagreements of another sort can plunge it into confusion. The body of scientific knowledge cannot develop without agreement over common standards amongst scientists. By the middle of the nineteenth century the emerging science of chemistry found itself in serious difficulties. No one could agree over a common international system for measuring the weights of the different elements.

The atoms of different elements were of course too small to be weighed individually. Everyone understood that their weight could be determined only on a relative basis. So in line with Dalton's suggestion, it had been agreed that all elements should be weighed relative to the lightest known element, hydrogen, which would be assigned a weight of one unit. But how should these relative weights be calculated? One school of thought favoured the atomic weight method. This relied upon Amedeo Avogadro's hypothesis that equal volumes of gases under similar temperature and pressure contained equal numbers of molecules. Thus all that was required was to weigh one volume against a similar volume of hydrogen.

The other school of thought favoured the equivalent weight method. This measured the weight of an element according to the relative amount which reacted chemically with a single amount of hydrogen, or a calculable equivalent. The only trouble was, the atomic weights and the equivalent weights of elements proved different. For instance, the atomic weight of oxygen was 16, but its equivalent weight was 8. Calculations in chemical papers were more and more frequently using figures without indicating whether they were atomic or equivalent weights. The result was growing confusion – to say nothing of the growing danger at the laboratory level.

In September 1860 the first ever international chemistry congress was called at Karlsruhe in Germany to thrash out this matter. The congress attracted leading chemists from all over Europe, as well as many who had yet to make their name, like Mendeleyev. The future of chemistry depended upon the outcome.

The case for the atomic weight lobby was put by the fiery and charismatic Italian chemist Stanislao Cannizzaro. Besides being a great chemist Cannizzaro was also a great revolutionary. Just four months prior to the Karlsruhe congress he had forsaken his post as professor of chemistry at Genoa to join Garibaldi's Red Shirts at Palermo. Cannizzaro's performance at Karlsruhe was in a similarly heroic vein. In ringing tones, he pointed out to the assembled delegates that the equivalent weight method was based on a ruinous misapprehension. The reason why the equivalent weight of oxygen was half the atomic weight was as follows: one volume of oxygen gas molecules weighed 16 times more than one volume of hydrogen gas molecules. But this same one volume of hydrogen gas molecules (H_2) reacted with only 8 volumes of oxygen gas molecules (O_2) to form water (H_2O). Similar circumstances also affected other elements.

Mendeleyev had never before heard science delivered with such fervour, and was completely overwhelmed. 'I vividly remember the impression produced by his speeches, which admitted of no compromise and seemed to advocate truth itself . . . The ideas of Cannizzaro [were] the only ones which could stand criticism and which represented the atom as "the smallest portion of an element which enters into a molecule or its compound". Only such real atomic weights . . . could afford a basis for generalization.' Mendeleyev's understanding of the nature of the atom and its atomic weight underwent a significant deepening. Without this crucial notion of atomic weight, any prospect of discovering a pattern amongst the properties of the elements would have been out of the question.

In 1861 Mendeleyev returned to St Petersburg. This was to be one of the most significant dates in Russian history: Tsar Alexander II finally decreed the liberation of the serfs. (Wilberforce's bill for the abolition of slavery had passed through the British Parliament in 1833, though the problem of slavery in the southern states of American was not tackled until the outbreak of the Civil War in 1861.) Tens of millions of Russians ceased at the stroke of a pen to be mere property, chattels shackled to their masters' estates. The liberated serfs, as well as their illiberal masters, were left in a state of bewilderment. No one knew what to do.

The teaching of chemistry in Russia was in a similarly bereft and bewildered state. On his return to St Petersburg, Mendeleyev took up a teaching post at the Technical Institute. To his astonishment he found that Russia simply hadn't heard of the fundamental advances in modern chemistry which were taking place all over Europe. (The development of spectroscopy, the discovery of new elements, the debate on atomic weight – these were but a few of the major events which had taken place during his absence.) The new enthusiastic young chemistry lecturer, who delivered lectures about the latest exciting developments in Europe, soon began attracting attention. His intense, deep-set blue eyes, his flowing beard and flowing hair made him an almost messianic figure. But this was a new kind of messiah in Russia: a prophet of science and rationalism. And when Mendeleyev discovered that there just wasn't such a thing as a Russian textbook on modern organic chemistry, he sat down and wrote one – completing five hundred pages in just sixty days. Mendeleyev was beginning to gain a name for himself, and was soon earning good money.

In 1864 he became a professor, and a year later he was able to take out a lease on a small estate two hundred miles south-east of St Petersburg at Tver. The liberation of the serfs meant that many landowners saw the value of their property collapse. Following the example of his mother, Mendeleyev set up a humanist programme for the peasants on his estate, at the same time introducing scientific farming methods. When the success of these methods became apparent, he began to receive visits from neighbouring peasants. 'Do you have a magic charm, or have you got a real knack for this sort of thing?' one wary peasant inquired of his new townee neighbour. Mendeleyev passed on the secret of his 'knack', and cooperatives all over the province were soon relying on his advice for such things as cheese-making and crop yields.

Mendeleyev's evangelism for the practical benefits of science was viewed with equal amazement by the authorities, who retained an orthodox medieval approach to such miraculous matters. By 1867 Mendeleyev was being sent as far afield as Baku in the Caucasus (to advise on the establishment of an oil industry) and Paris (to organize the Russian pavilion at the Exposition Internationale).

Meanwhile, at the age of twenty-nine Mendeleyev had got married. The marriage was to produce one son and one daughter, but was not otherwise a success. Mendeleyev's need to be mothered and spared contradiction on any point (the mascot malady), together with his foul temper (Rumpelstiltskin complex) and his propensity to retire to his study for superhuman bouts of work (the vanishing-man syndrome) meant that he was impossible to live with in any normal sense of the word. Fortunately his wife proved an imaginative and resourceful woman. She wisely chose to spend her time on the estate at Tver, except when her husband arrived there from St Petersburg, when she and the children would depart for the Mendeleyev town residence. In this way the marriage managed to survive, without the cohabitation which is the ruin of so many relationships.

At the age of thirty-two Mendeleyev was appointed professor of general chemistry at the University of St Petersburg, an exceptionally prestigious post for one so young. Here his senior colleague was Russia's finest chemist, Aleksandr Butlerov, who did pioneering work on the structure of chemical compounds. Sadly Butlerov was to succumb to the prevalent Russian disease of spiritualism. In the interests of science Mendeleyev was called upon to investigate Butlerov's supra-chemical claims, which he found to be nonsense. But it seems they remained friends despite this spectral disagreement.

Mendeleyev's enthusiastic conviction and eccentric appearance made him an immediate hit with the students. Even those who knew little about chemistry were able to appreciate the nuggets of chemical lore, anecdotal information and encyclopedic asides (on everything from astrophysics to zoology, astronomy to zymosis) with which he peppered his lectures. Amongst his students was the future anarchist leader Prince Kropotkin, who recalled: 'The hall was always crowded with something like two hundred students, many of whom, I am afraid, could not follow Mendeleyev, but for the few of us who could it was a stimulant to the intellect and a lesson in scientific thinking which must have left deep traces in their development, as it did in mine.' Leaving aside Mendeleyev's unwitting role as an intellectual stimulant to the development of anarchism, it is evident that he had an inspirational effect on his students. One of his more perceptive enthusiasts remarked: 'Thanks to Mendeleyev I began to perceive that chemistry was really a science.' In the eyes of many, chemistry had yet to come of age. It still appeared as little more than a compendium of technical know-how and unrelated facts – with no overall guiding principle.

Mendeleyev was highly conscious of this defect in his chosen subject. His chemistry lectures may have included miscellaneous asides on everything under the sun, but his guiding intellectual principle was synthesis. His mind was constitutionally averse to leaving a fact to stand by itself. As he put it: 'The edifice of science requires not only material, but also a plan, and necessitates the work of preparing the materials, putting them together, working out the plans and symmetrical proportions of the various parts.' For Mendeleyev this was more than just an intellectual task: 'To conceive, understand, and grasp the whole symmetry of the scientific edifice, including its unfinished portions, is equivalent to tasting that enjoyment only conveyed by the highest forms of beauty and truth.' He was interested in uncovering 'the philosophical principles of our science which form its fundamental theme'.

Noble sentiments indeed. Yet Mendeleyev had not forgotten the antipathies of his youth. Despite the sentiments quoted above, he remained implacably opposed to the classics and to philosophy itself. In his view these bred only 'self-deceit, illusion, presumption and selfishness'. The way he saw it: 'classicists are only fit to be landowners, capitalists, civil servants, men of letters, critics . . . We could live at the present day without a Plato, but a double number of Newtons is required to discover the secrets of nature, to bring life into harmony with the laws of nature.'

Mendeleyev's philistinism (or at best anti-culturalism) was perhaps pardonable in view of the state of Russian education during this period. Yet it remains an archetypical example of the arrogance of science, an attitude which has persisted to this day. The American quantum physicist Richard Feynman frequently expressed his disdain for 'culture', and regarded philosophy as a useless irrelevance. Such views remain widespread, but are put in their place by Erwin Schrödinger, of 'Schrödinger's cat' fame, an Austrian of considerable culture as well as one of the twentieth century's finest scientific minds: 'Science is necessary, but not sufficient, in establishing one's view of the world.' The French philosopher Comte had in fact been right when he claimed that there are things which science will never know. It was merely his examples that proved to be wrong.

Once again, Mendeleyev found himself up against the problem of Russia's backwardness. His lectures were on inorganic chemistry – but his students were hampered by the fact that there was simply no adequate textbook on this subject available in Russian. Having written the only available textbook on modern organic chemistry in record time, Mendeleyev now sat down to write the definitive work on inorganic chemistry.

Then, much as now, organic chemistry covered compounds that formed the basis of living matter – largely substances combining carbon with hydrogen, oxygen or nitrogen. Inorganic chemistry covered the dead remainder: the study of the properties of the basic chemical elements and their compounds.

By early 1869 Mendeleyev had completed the first volume of his projected two-volume *The Principles of Chemistry*. This was to be his masterpiece: the finest chemistry textbook of its era. It would be translated into all major languages and remain a standard work until the early years of the twentieth century. Yet this was no stuffy orthodox textbook. Its style was very much in the manner of the man. Like his lectures, the text was peppered with footnote asides on a huge variety of subjects, chemical anecdotes and scientific speculations. Indeed, the footnotes were longer than the actual text. And even the latter contained many a fine nugget. Not for nothing did Mendeleyev claim *The Principles* as 'my favourite child', adding that it contained 'my likeness, my experiences . . . and my most sincere scientific ideas'.

Despite its likeness to its author, *The Principles of Chemistry* is no work of woolly eccentricity. This goldmine of information had a very definite structure. The elements, and their compounds, were treated together in groups with similar properties, each following on from the previous ones. For instance, the end of the first volume covered the halogen group, consisting of fluorine, chlorine, bromine and iodine. This group took its name from the Greek *halos* ('salt') and *-gen*. Each of the halogen group of elements combined with sodium to produce salts which had very similar properties, the best known of these being of course table salt – sodium chloride. The halogens also combined readily with potassium. So the obvious logical step was to start volume two with the alkali metals group, which contained sodium and potassium. This Mendeleyev reckoned would occupy the first two chapters.

By the morning of Friday 14 February 1869 these two chapters were complete. Mendeleyev now faced the pressing problem of what group of elements to deal with next. The structure of his entire book depended upon this. It was necessary to discover some underlying principle by which the elements could be ordered. And there was no time to lose: he had to get this done over the weekend. On Monday he was due to catch the train for Tver, where he had to deliver a speech to a delegation of cheese-makers, followed by a three-day tour of inspection of the local farms . . . If he could just sort out this problem of the elements, he would be able to start straight back into writing his textbook as soon as he returned from the country.

Mendeleyev sat at his littered desk amidst the clutter of his study: a wild-haired, gnome-like figure obsessively combing the fingers of his left hand through the points of his long straggly beard. From the dimness above his head the portraits of Galileo, Descartes, Newton and Faraday stared down as his pen scratched away amidst the encroaching disorder of spilling papers, books and obscure mechanical devices.

Painstakingly Mendeleyev began sifting through his encyclopedic knowledge of the chemical elements, in search of some pattern of properties which might link the groups of similar elements. There had to be a key to it all somewhere. The elements couldn't just have a random set of properties: that would be unscientific. Admittedly, they had a definite order of atomic weights, but that was surely only part of it. That just accounted for the physical properties. What about an order amongst the chemical properties? De Chancourtois claimed to have discovered some kind of recurring pattern, but it was impossible from his paper to work out precisely what this pattern was. And even de Chancourtois admitted that it didn't seem to fit all the elements. He kept referring to some mysterious 'telluric screw' which was never fully explained. It seemed obvious that de Chancourtois himself didn't really understand what he was trying to describe. Yet whatever it was he had glimpsed had evidently convinced him there was a pattern of some kind. Mendeleyev felt certain that de Chancourtois was right. The entire weight of his chemical knowledge inclined him in favour of this intuition. Again and again he racked his brains. Again and again he tried building up the elements one by one to form some kind of structure. But each time the structure tumbled to the ground like a pack of cards. By the morning of Monday 17 February Mendeleyev still hadn't come up with anything.

That morning at breakfast Mendeleyev looked through the letters which had arrived in the post. Amongst these was a communication from the secretary of the Voluntary Economic Cooperative of Tver, two hundred miles south-east of St Petersburg. This gave the details of his forthcoming meeting with the cheese-makers, which had been arranged for that afternoon. The following three days had been set aside for a tour of inspection of the province's cheese-making centres, just as he had requested. All the arrangements had been made in preparation for his visit.

Mendeleyev was due to catch the train from St Petersburg's Moscow Station immediately after breakfast. His wooden travelling trunk was standing packed and ready by the front door. Outside, the horse-drawn sleigh was waiting in the snow-covered street.

Mendeleyev eventually took up the letter from the breakfast table, along with his mug of tea, and retired to his study. Here he sat down once more at his desk. He turned over the letter he had been carrying, set down his mug of tea and began jotting some notes on the back of the letter. We know these details, because this letter from the secretary of the Voluntary Economic Cooperative of Tver has been preserved, and still bears the ring-mark of Mendeleyev's mug. The notes on the back of the letter list a number of elements in order of their atomic weight. Mendeleyev evidently felt certain that the key lay in this natural and obvious clue – the ascending order of weights. Here undeniably was one order. The trouble was, it didn't appear to explain anything: just that one element was heavier than another. So what about the groups of elements with similar properties, such as the halogens: fluorine (F), chlorine (Cl), bromine (Br) and iodine (I)? But the atomic weights of the halogen group differed widely:

F = 19   Cl = 35   Br = 80   I = 127

It was the same with the oxygen group of elements: oxygen (O), sulphur (S), selenium (Se) and tellurium (Te). These too exhibited significantly similar chemical properties, but their weights differed as follows:

O = 16   S = 32   Se = 79   Te = 128

And the nitrogen group of elements, nitrogen (N), phosphorus

(P), arsenic (As) and antinomy (Sb), were just as disparate when it came to their atomic weights:

N = 14 P = 41 As = 75 Sb = 122

But when Mendeleyev listed these three groups above one another on the back of the letter, he noticed that a pattern began to emerge:

F = 19	Cl = 35	Br = 80	I = 127
O = 16	S = 32	Se = 79	Te = 128
N = 14	P = 41	As = 75	Sb = 122

Reading from the foot of the vertical columns, the elements ascended in atomic weight. Only phosphorus (P) in the second column, and tellurium (Te) in the fourth column didn't fit. All the others fell into place.

What did this mean? It didn't appear to make any sense. But surely this couldn't be a mere coincidence? Even so, it only accounted for a dozen elements. There were still over fifty left, and they remained as disordered as before.

According to Mendeleyev's friend A. A. Inostrantzev, who visited him that day, Mendeleyev had by now been working on this problem for three days and three nights without ceasing. As we know, Mendeleyev was capable of prodigious feats of concentration when the mood took him – such as when he'd written his five-hundred-page textbook on organic chemistry in just sixty days. Mendeleyev began scribbling out lists of the other elements, aware by now that time was running out. He still had to catch the train for Tver. The handwriting in his notes, and the frequent crossings-out, betray his increasing agitation. He was convinced he was on the right track. He had already had one inkling, but somehow he just couldn't see beyond it.

It must have been at this point that Mendeleyev had his brainwave – making the inspired connection between the problem of the elements and his favourite card game, patience. He began writing out the names of the elements on a series of blank cards, adding their atomic weights and chemical properties.

Presumably it was around now that he went to the study door and ordered his servant to dismiss the sleigh-driver, telling him to come back in time for the afternoon train. As the tinkling bells of the departing sleigh faded into the snow-padded silence beyond the window, Mendeleyev began concentrating on the sea of cards spread out before him over the desk.

Mendeleyev records in an earlier diary how sometimes after making a first step towards a discovery he would become elated. But later, if he was unable to follow through this initial insight, he would lapse into a state of deep depression, sometimes finding himself reduced to tears.

It was in such a depressed state that he was found by his friend Inostrantzev, when he called round to see Mendeleyev on the afternoon of 17 February. Mendeleyev knew he was on the verge of a great discovery, but he just couldn't grasp it. 'It's all formed in my head,' he complained bitterly, 'but I can't express it.' Exhausted, he rested his shaggy head in his hands.

Passing over the dubious philosophy implicit in this remark, one can certainly sense Mendeleyev's frustration.

Once again it was getting late. The afternoon train would soon be departing from the Moscow Station for Tver. And once again the sleigh had appeared in the street outside, its driver hunched against the chill of the fading early afternoon light. There was simply no time to explore every avenue of possibility amongst the sea of cards spread before him – no time for meditating on the pros and cons of each alternative scheme. He had to act quickly: guess, rely on his intuition, decide. Fortunately Mendeleyev's intuition was one of his main strengths.

What Mendeleyev had noticed was the similarity between the elements and the game of patience. In patience the cards had to be aligned according to suit and descending numerical order:

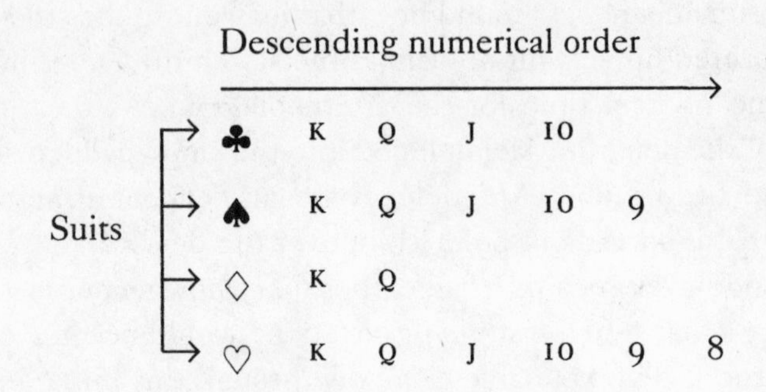

What Mendeleyev was looking for amongst the elements appeared to be something very similar: a pattern listing the elements according to groups of similar properties (like the suits), with the elements in each group aligned in sequence of their atomic weights (echoing the sequence of numerical order in the suits):

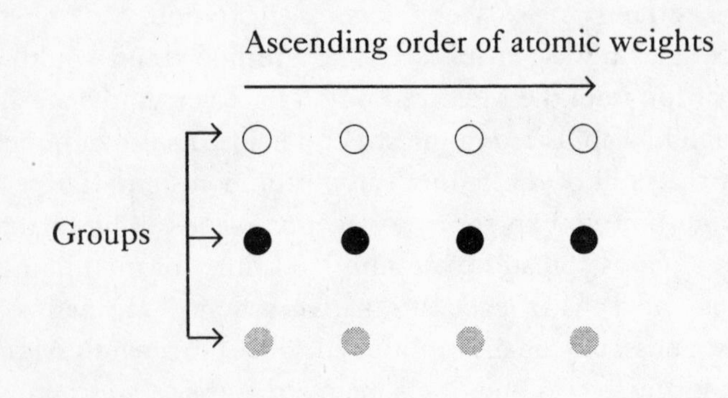

This game of 'chemical patience', as he called it, evidently confirmed, and seemed to point beyond, Mendeleyev's initial insight – concerning the emergent, but nonetheless flawed, pattern amongst the halogen, oxygen and nitrogen groups. Everything seemed to be coming together.

By now the fading dusk outside the window had given way to the frigid darkness of night. Despite his growing exhaustion, he knew he couldn't stop now. Mendeleyev was possessed by the feeling that he was on the brink of a momentous discovery.

So it comes as something of a disappointment to record that at this point he was overcome with fatigue. He leaned forward, resting his head on his arms amidst the scattered cards on his desktop. Almost immediately he fell asleep, and had a dream.

14

The Periodic Table

In Mendeleyev's own words: 'I saw in a dream a table where all the elements fell into place as required. Awakening, I immediately wrote it down on a piece of paper.' In his dream, Mendeleyev had realized that when the elements were listed in order of their atomic weights, their properties repeated in a series of periodic intervals. For this reason, he named his discovery the Periodic Table of the Elements.

On the opposite page is the design of the Periodic Table which Mendeleyev published two weeks later in his historic paper 'A Suggested System of the Elements'. Reading from the top of the furthest left column, the vertical columns list the elements in order of ascending atomic weights. The horizontal rows list the elements in groups with similar graduating properties.

As can be seen, the second vertical column resembles Newlands's law of octaves, but this hardly repeats in the elements of higher atomic weight. Similarly, the partial patterns of Döbereiner and de Chancourtois are also accounted for. Mendeleyev's Periodic Table followed a less rigid pattern, yet this pattern seemed to encompass all the known elements.

However, even Mendeleyev had to concede that at first sight there appeared to be a number of anomalies in this pattern. For a start, if all the elements were grouped horizontally according to their properties, this meant that some of their atomic weights did not conform to the precise ascending order: for instance, at the foot of the third vertical column, thorium (Th = 118). In such

ОПЫТЪ СИСТЕМЫ ЭЛЕМЕНТОВЪ.

ОСНОВАННОЙ НА ИХЪ АТОМНОМЪ ВѢСѢ И ХИМИЧЕСКОМЪ СХОДСТВѢ.

			Ti=50	Zr=90	?=180.
			V=51	Nb=94	Ta=182.
			Cr=52	Mo=96	W=186.
			Mn=55	Rh=104,4	Pt=197,4.
			Fe=56	Rn=104,4	Ir=198.
			Ni=Co=59	Pl=106,6	Os=199.
	H=1		Cu=63,4	Ag=108	Hg=200.
	Be=9,4	Mg=24	Zn=65,2	Cd=112	
	B=11	Al=27,4	?=68	Ur=116	Au=197?
	C=12	Si=28	?=70	Sn=118	
	N=14	P=31	As=75	Sb=122	Bi=210?
	O=16	S=32	Se=79,4	Te=128?	
	F=19	Cl=35,5	Br=80	I=127	
Li=7	Na=23	K=39	Rb=85,4	Cs=133	Tl=204.
		Ca=40	Sr=87,6	Ba=137	Pb=207.
		?=45	Ce=92		
		?Er=56	La=94		
		?Yt=60	Di=95		
		?In=75,6	Th=118?		

Д. Менделѣевъ

cases Mendeleyev questioned the atomic weight of the element, suggesting that it had been calculated incorrectly. Here, he arrogantly claimed, science was wrong and he was right! Even more bold was his suggestion to account for other anomalies in his Periodic Table. Where no element fitted into the pattern, he simply left a gap. He predicted that these gaps would one day be filled by elements which had not yet been discovered. For instance, in the ninth horizontal row (the bismuth group beginning B = 11), he predicted that there was a hitherto unknown element between aluminium (Al = 27.1) and uranium (Ur = 116). This element he named eka-aluminium, predicting that when it was discovered its atomic weight would be 68. He even went so far as to predict its properties, which would lie between those of aluminium and uranium. Similarly, in the next horizontal line down, the carbon group beginning C = 12, he predicted a further element between silicon (S = 28) and tin (Sn = 118), which he marked as ? = 70. This he called eka-silicon, and likewise described its probable properties.

Despite these apparent anomalies in his Periodic Table, Mendeleyev felt sure that he was right. A further piece of evidence only confirmed him in his view. The pattern revealed in his Periodic Table was uncannily echoed by a pattern in the sequence of valencies exhibited by each element, i.e. in the measure of its atoms' ability to combine with other atoms. For instance, lithium (Li = 7 in the table) had a valency of 1. That is, if the atom was a ball it had one 'arm' enabling it to join with other atoms. Next in the sequence of atomic weights came beryllium (Be = 9.4), which had a valency of 2, enabling it to join with two other atoms. The following element, boron (B = 11), had a valency of 3; then came carbon (C = 12) with 4. Then the sequence fell, so that the overall order read: 1, 2, 3, 4, 3, 2, 1. This periodic rise and fall was more or less repeated along the entire sequence of atomic weights. But when the elements were arranged in horizontal groups of similar properties, as he had arranged them in his Periodic Table, the elements in the same groups tended to have the same valency. Thus the elements in the nitrogen group (eleventh horizontal line from the top, beginning N = 14) had a valency of 3. The next group down, the oxygen group (beginning O = 16), had a valency of 2; and the group below had a valency of 1. Again, there were a number of discrepancies where the valencies didn't quite appear to fit the pattern, or elements had to be rearranged out of order, but Mendeleyev felt sure that these anomalies too were susceptible to explanation. He remained convinced that his Periodic Law, as he called it, was the answer. As he was later to remark: 'Although I have had my doubts about some obscure points, yet I have never once doubted the universality of this law, because it could not possibly be the result of chance.'

Others remained less than convinced. This so-called 'law' was typical of Russian science: it had none of the rigour of its Western counterparts. Quite simply, the Periodic Table had far too many holes in it. How could Mendeleyev rely upon certain atomic weights having been miscalculated? Whoever heard of a scientific theory which relied upon scientific errors?

But Mendeleyev's Periodic Table was soon to receive support in a manner which he least expected. The German scientist Julius Meyer published a paper claiming that he had discovered the Periodic Table. This was surely more than just a misguided coincidence.

The lives of Mendeleyev and Meyer in fact contained several coincidences. A few years later than Mendeleyev, Meyer too had studied chemistry at Heidelberg with Bunsen and Kirchhoff, the renowned inventors of spectroscopy who had even managed to discover new elements on the surface of the sun. Though unlike Mendeleyev, Meyer had not stormed out of Bunsen's laboratory. As a result he had acquired a profound insight into the nature of the elements (rather than obstinately passing his time in a home lab testing the solubility of alcohol in water). Meyer too had attended the famous Karlsruhe conference in 1860, and like Mendeleyev had been inspired by Cannizzaro's fiery speech in favour of atomic weights.

Working along similar lines to Mendeleyev, Meyer had eventually discovered an almost identical pattern amongst the elements, at precisely the same time as Mendeleyev. So why is Mendeleyev given credit for the discovery of the Periodic Table? For a start, because he published his paper on the subject on 1 March 1869, just two weeks after his initial discovery – whereas Meyer didn't publish until the following year. But decisively, Meyer's conclusions were more tentative. He couldn't fully account for the anomalies in his table – the elements which were out of order, those which didn't appear to fit into their apparent group and the glaring gaps. When critics pointed out the discrepancies between Meyer's 'law' and the facts, Meyer had no defence. Mendeleyev, on the other hand, took the offensive. He was willing to back his chemical insight in the face of all the 'facts'.

Understandably, the scientific world remained unconvinced. How could you base a scientific law upon discoveries which had not yet been made? This reliance upon undiscovered chemical elements was sheer fantasy. Mendelelev's position began to look increasingly shaky as time passed and still no scientific evidence emerged to back up his wild assertions. No new elements were discovered with the properties of 'eka-aluminium' or 'eka-silicon'. Indeed, the years immediately following 1869 were to prove surprisingly barren where the discovery of new elements was concerned. Then in the late summer of 1874 the Académie des Sciences in Paris received a dramatic letter from the French chemist Paul Lecoq de Boisbaudran in which he announced: 'During the night before last, on 27 August 1875 between three and four in the morning, I discovered a new element in a sample of zinc sulphide from the Pierrefitte mine in the Pyrenees.' This element he named gallium, after the Latin for France, Gallia. (Though it has been suggested that Lecoq's motives might not have been quite so patriotic and selfless as they appeared. *Le coq* in French means 'cockerel', which tranlates into Latin as *gallus*.)

Lecoq's newly discovered element was found to have an atomic weight of 69, and its properties indicated that it belonged to the boron group, between aluminium and uranium. The new element gallium all but exactly matched the properties which Mendeleyev had predicted for eka-aluminium. But when Lecoq calculated the specific gravity of gallium, he found it to be 4.7 – whereas Mendeleyev had predicted that eka-aluminium would have a specific gravity of 5.9. Here was a glaring discrepancy which could not be overlooked. Was it possible that Mendeleyev's other 'predictions' had been nothing more than a series of lucky guesses?

As soon as Mendeleyev heard that Lecoq's laboratory finding did not match with his theoretical prediction, he reacted in characteristic fashion. Immediately he dispatched a letter to Lecoq informing him that his sample of gallium was insufficiently pure, suggesting he repeat the experiment with another sample. Lecoq dutifully repeated his experiment with a larger sample, which he subjected to rigorous purification. And this time he found that the specific gravity of gallium was 5.9, just as Mendeleyev had predicted!

Five years later came confirmation that this was certainly no fluke. During routine analysis of the mineral argyrodite, which had recently been discovered in a mine near Freiburg, the German chemist Clemens Winkler detected the presence of a hitherto undiscovered element. This he named germanium (after his native land). In the course of various minor revisions to his Periodic Table, Mendeleyev had calculated that eka-silicon would not have an atomic weight of 70, but one closer to 72 – yet his predictions concerning its other features remained unaltered. It would be a dark grey metallic element with properties between those of silicon and tin. It would have a specific gravity of 5.5, its oxide would have a specific gravity of 4.7 and its compound with chlorine would have a specific gravity of 1.9. Winkler found germanium to be a grey substance with a metallic glint, with an atomic weight of 72.73. It had a specific gravity of 5.47, its oxide had a specific gravity of 4.7 and its chloride had a specific gravity of 1.887. No one could now doubt Mendeleyev's Periodic Law.

With the Periodic Table chemistry came of age. Like the axioms of geometry, Newtonian physics and Darwinian biology, chemistry now had a central idea upon which an entire new range of science could be built. Mendeleyev had classified the building blocks of the universe.

Mendeleyev felt sure that such a discovery would lead to far-reaching advances in science. He speculated that in future centuries his Periodic Table might perhaps indicate the origins of the universe, the pattern upon which life itself was based or even the ultimate secret of matter. This was to happen, but not at all in the way he expected. Even during Mendeleyev's lifetime it was discovered that certain elements in the Periodic Table were liable to decay. Mendeleyev could not accept this: to him the Periodic Table was an absolute. However, it was the very position of these decaying elements in his table which led scientists to understand precisely what was happening. The atom was not the ultimate particle. Nuclear physics was born. In its turn, nuclear physics would come up with its own collection of unique subnuclear particles. Was it possible that these too might conform to a pattern similar to the Periodic Table of the elements? In 1981 the American physicist Murray Gell-Mann, inspired by Mendeleyev's example, came up with a classification table for subatomic particles, which he named the eightfold way. These grouped the particles in families exhibiting similar properties, in a manner which echoed the method Mendeleyev had first used to group the elements. But modern science moves fast. Now the eightfold way has suffered the same fate as Mendeleyev's Periodic Table. The particles it classifies have been found not to be absolute. They seem to consist of even more minute entities known as superstrings.

However, despite these far-reaching advances, the Periodic Table first discovered by Mendeleyev remains the basis of modern chemistry. It has been used to predict the possible properties of all manner of molecular combinations of atomic elements. This is of particular use in the synthesis of complex new drugs. Similarly, the precise knowledge of each different atomic element's ability to combine with other atoms has led to the most spectacular advances in chemistry. Our understanding of the constitution of the immensely complex DNA molecule, 'the pattern of life', would not have been possible without such knowledge.

Mendeleyev's hunch that his Periodic Table might assist in the discovery of the origins of the universe has also proved justified. Cosmologists speculating on what happened in the seconds after the Big Bang find themselves working forwards from the foundation of nuclear particles (in the first three seconds) to the formation of the first atoms (one million years later). How these first simple atoms became the complex structure of the Periodic Table is the secret of the evolution of the universe. We know how it began; Mendeleyev showed us how it is at present. What happened in between is the vital episode which we are only now beginning to piece together.

The Periodic Table of the elements has undergone several adjustments and rearrangements during the century or so since Mendeleyev's original discovery. Yet the modern versions of the Periodic Table (of which there are several) are all still incontrovertibly based upon his core structure. This has been able to take on board almost double the number of elements for which it originally accounted, including an entirely new group, and several subsequent regroupings of elements. The properties, valencies and weights of elements are now known to result from the arrangement of subatomic particles within the atom. Yet nuclear physics has mostly confirmed Mendeleyev's original hunches concerning atomic weights, missing elements and their properties. It offers an overall explanation based on the most sophisticated experimental evidence, where Mendeleyev could only speculate. What Mendeleyev discovered on 17 February 1869 was the culmination of a two-and-a-half-thousand-year epic: a wayward parable of human aspiration.

In 1955 element 101 was discovered and duly took its place in the Periodic Table. It was named mendelevium, in recognition of Mendeleyev's supreme achievement. Appropriately, it is an unstable element, liable to spontaneous nuclear fission.

Further Reading

Because this is intended as a popular book, I have not included an exhaustive list of sources. Quotes in the text are mostly attributed, and many relevant works are mentioned. Listed below are sources I have used for each chapter which may prove of interest for further reading.

A word of warning when searching for further material: Mendeleyev's name is westernized from Russian Cyrillic script. This has resulted in a variety of different spellings. The variant I have used seems the most logical for English, and is the one generally recognized. However, this matter has not been formalized and the following variants may also be found (in titles, indexes, papers, etc.): Mendeleev (which is how he signed himself in English), Mendeléev, Mendeléïev, Mendeleïeff, Mendeleyeff and other variants.

Prologue

Dmitri Mendeléïev, by Paul Kolodkine, Paris, Seghers, 1961. In French. One of the few biographies not in Russian.

'On the Question of the Psychology of Scientific Creativity (On the Occasion of the Discovery by D. I. Mendeleev of the Periodic Law)', by B. M. Kedrov, in *Soviet Review*, Vol. 8, 2 (1967), pp. 26–45. A detailed investigation of what happened, translated from the Russian.

'Factors Which Led Mendeleev to the Periodic Law', by Henry M. Leicester, in *Chymia*, Vol. 1 (1948).

1. *In the Beginning*

Lives of Eminent Philosophers, by Diogenes Laertius, London, Heineman, 1980. The major source on the early philosophers. Highly readable, but not always reliable.

A History of Western Philosophy, Vol. 1, *The Classical Mind*, by W. T. Jones, New York, Harcourt Brace, 1970. Particularly good on ideas and the scientific aspect of the early philosophers.

The Discovery of the Elements, by M. E. Weeks and H. M. Leicester, Journal of Chemical Education (US), 1968. The exhaustive work, which runs to almost five hundred pages. A mine of fascinating lore, which I have consulted throughout.

2. *The Practice of Alchemy*

Through Alchemy to Chemistry, by John Read, London, Bell, 1957.

The Origins of Alchemy, by Jack Linsay, London, Muller, 1970.

Avicenna: His Life and Works, by Soheil M. Afnan, London, Allen & Unwin, 1958. A fascinating biography, making full use of what little reliable material is available.

3. *Genius and Gibberish*

History of Chemistry, by J. R. Partington, London, Macmillan, 1962. The four-volume early standard work; I have consulted it throughout.

Dictionary of Scientific Biography, ed. C. C. Gillespie, New York, Scribner's, 1974. The sixteen-volume definitive biographical reference work, which I have also consulted throughout.

Alchemy and Mysticism, by Alexander Roob, London, Taschen, 1998. Seven hundred pages filled with illustrations: good for background and nuggets of information.

4. *Paracelsus*

Paracelsus: Magic into Science, by Henry M. Pachter, New York, Schuman, 1971. The best biography of Paracelsus available in English.

Crucibles: The Story of Chemistry, by Bernard Jaffe, Dover, 1998. Contains a good brief biographical chapter on Paracelsus, as well as chapters on many other leading figures in the history of chemistry.

The longish entry in Volume 10 of Gillespie's *Dictionary of Scientific Biography* (see above under Chapter 3) is particularly informative.

5. *Trial and Error*

The Rainbow: From Myth to Mathematics, by Carl B. Boyer, New York, 1959. Places Dietrich von Freiburg in the larger context, and unravels the whole story.

For further details of Nicholas of Cusa, see Gillespie, op. cit., Vol. 3.

Nicholas Copernicus, by Fred Hoyle, London, Heinemann, 1973. Essay by a leading contemporary astronomer, with interesting professional insights.

Giordano Bruno: His Life and Thought, by Dorothy Waley Singer, New York, Schuman, 1950. Still probably the best general biography.

6. *The Elements of Science*

Galileo: A Life, by James Reston jnr, London, Cassell, 1994. The latest of many fine biographies: comparatively short, but clear and to the point.

Descartes: An Intellectual Biography, by Stephen Gaukroger, Oxford, OUP, 1995. A useful exploration of his philosophy and science which, despite the title, also contains details of his life.

Francis Bacon: The History of a Character Assassination, by Nieves Mathews, New Haven, Yale UP, 1996. A useful corrective to many of the myths that have built up around Bacon's name.

A Historical Introduction to the Philosophy of Science, by John Losee, Oxford, OUP, 1993. The best comprehensive guide to this often perplexing subject.

7. *A Born-again Science*

The Life of the Honourable Robert Boyle, by R. E. W. Maddison, London, Taylor & Francis, 1969. Good comprehensive biography with much fascinating period detail.

Isaac Newton: The Last Sorcerer, by Michael White, London, Fourth Estate, 1997. Recent biography which focuses on the more controversial alchemical aspects of Newton's work.

Van Helmont: Alchemist, Physician, Philosopher, by H. Stanley Redgrove and I. M. L. Redgrove, London, William Rider, 1922. A short, rare work, but practically the only information available in English.

8. *Things Never Seen Before*

Weeks and Leicester, op. cit. The exhaustive work on the subject, also includes details on the lives of the discoverers.

Man and the Chemical Elements, by J. Newton Friend, London, Charles Graham, 1951. Highly readable history of the discovery of the elements and their discoverers.

There is no English biography of Karl Scheele, but there is a good outline of his life and work in *Dictionary of Scientific Biography*, op.cit., Vol. 12, pp. 143–50.

The Chemical Elements, by I. Nechaev and Gerald Jenkins, Diss (Norfolk), Tarquin, 1997. Excellent short popular survey of the chemical elements and much other chemical lore.

9. *The Great Phlogiston Mystery*

Unfortunately there is no full-scale biography of Becher available in English, but there is a good summary of his remarkable life in Jaffe, op. cit. (under chapter 4).

Likewise there is no readily available biography of Stahl, but you can find a brief chapter on his life in *Great Chemists*, ed. Eduard Farber, Interscience, 1961.

Joseph Priestley by F. W. Gibbs, London, Nelson, 1965. The best popular biography of a remarkable man of science and principle.

Cavendish, by Christina Jungnickel and Russell McCormmach, American Philosophical Society, 1996. Good on the exceptional science as well as the exceptional man: many fine illustrations.

10. *The Mystery Solved*

Lavoisier: Chemist, Biologist, Economist, by Jean-Pierre Poirier, University of Pennsylvania Press, 1996. Full-scale biography, rich in detail of life, work and times. Translated from the French.

Lavoisier, by Ferenc Szabadvary, University of Cincinnati, 1977. A shorter, but highly readable account, translated from the Hungarian.

Fontana History of Chemistry, by William H. Brock, London, Fontana, 1992. The best available popular work on the subject, with good chapters on the phlogistonists and Lavoisier.

11. *A Formula for Chemistry*

John Dalton and the Atom, by Frank Greenaway, London, Heinemann, 1966. Readable and informative popular biography, good on both the science and the times.

Enlightenment Science in the Romantic Era, ed. Evan Melhado and Tore Frangsmyr, Cambridge, CUP, 1992. The chemistry of Berzelius and its cultural setting in early-nineteenth-century Europe.

12. *The Search for a Hidden Structure*

Virtually the only extended sources for Döbereiner and de Chancourtois are the appropriate sections in the *Dictionary of Scientific Biography*, op. cit.

Chemical Age, 59 (1948), contains an article by J. A. Cameron entitled 'J. A. R. Newlands (1837–1898), A Pioneer Whom Chemists Ridiculed'.

Partington's *History of Chemistry*, op. cit., Vol. 4, Part 4, traces in more detail the search for a pattern amongst the elements.

'The Development of the Periodic Table' by John Emsley, in *Interdisciplinary Science Reviews*, Vol. 12 (1987).

13. *Mendeleyev*

Eminent Russian Scientists, by Cicely Kodiyan, Delhi, Konark, 1992. Biographies of the leading scientists before and after Mendeleyev.

See also titles listed for the Prologue.

14. *The Periodic Table*

The Periodic System of the Elements: A History of the First Hundred Years, by J. W. van Spronsen, London, Elsevier, 1969. Outlines the developments which have taken place since Mendeleyev's time.

Graphic Representations of the Periodic System during One Hundred Years, by Edward G. Mazurs, University of Alabama, 1974. Describes how the picturing of the Periodic Table has changed since Mendeleyev's original versions.

Index